Managing High-Technology Companies

Managing High-Technology Companies

Henry E. Riggs

LIFETIME LEARNING PUBLICATIONS
Belmont, California

A division of Wadsworth, Inc.

London, Singapore, Sydney, Toronto, Mexico City

To Gayle

Designer: Gary Head Design
Copy Editor: Sylvia Stein
Illustrator: John Foster
Composition: Graphic Typesetting Service

Printed in the United States of America

1 2 3 4 5 6 7 8 9 10 — 86 85 84 83

Library of Congress Cataloging in Publication Data

Riggs, Henry.
  Managing high-technology companies.

  (Lifetime learning series in managing high technology; 1)
  Includes bibliographical references and index.
  1. High technology industries—Management.
I. Title. II. Series
HD62.37.R53 1983   621,381'71'068   83- 17525
ISBN 0-534-02720-2

Series Preface

These are the first books addressed specifically to managers of companies whose strategies are rooted in technology—that is, to managers of *high-technology companies* who are compelled to view new technology not as a threat but as an opportunity.

The importance of such companies to the growth and health of developed economies around the world is increasingly recognized by both governments and managers. Yet despite the attention paid to high-technology industries by both the popular press and governments, there has been little writing directed towards improving management decision-making on questions that are unique to the companies comprising these industries. Books and articles have focused on service, as distinct from manufacturing, industries, on small companies as distinct from large ones, but not until now on technology-based companies—large and small, service and product—as a distinctive unit.

Managing high-technology companies is not, of course, wholly different from managing companies in which technology plays no strategic role—but it is *significantly* different. Each function of a high-technology business—research, product development, production, marketing, finance, planning, human relations—must respond to the challenges inherent in a *rapidly changing technology*. Rapidly changing technology itself creates risk, but, fortunately, the other side of the risk coin is opportunity. High-technology companies are complex, but that complexity implies both more sources of competitive leverage and greater opportunities for the well-managed company to distinguish itself. For managers of high-technology companies to be effective (as well as their suppliers of products and services), they must understand the critical elements of their business, the connections between these elements, and the ways in which their companies relate to or function with other companies. To be effective, these managers must understand their own position in the context of the total company.

High-technology industries present unique challenges, but they also offer fun, excitement, and reward to ambitious and well-trained managers. This series can help you improve your skills and your understanding, and thus both your contributions to and rewards from your high-technology company.

Contents

Chapter 8   General Management and Personnel Policies in High-Technology Companies   253

Preface

Audience

This book is for people, especially engineers, who have or aspire to have responsibility for managing complex, technological, productive enterprises, companies in which technology—both product and process technology—is the key to business strategy.

My emphasis here is less on the management *of* engineers than on management *for* engineers. While my focus is upon the general manager, middle-level managers in high-technology companies will also gain insights into their own positions in the context of the total company.

Features

My purpose is to provide an overview of high-technology companies, to point up the important effects that technology has on *all* functions within the business, to focus on the relationships between functions, and to provide a framework for designing and directing the strategy of the high-technology company. Specifically, this book examines the following issues:

- The critical relationships between *engineering, production,* and *marketing*, a set of relationships that creates the complexity, challenges, and opportunities inherent in high-technology companies.

- The pressure for growth, the linkage between financial

returns and growth, and the financing alternatives and constraints present in high-technology companies.

- The importance of quality in high-technology industries— quality of products, quality of processes, and quality of human life.

- The challenge of attracting, assimilating, motivating, stimulating, and managing creative staff members that are essential to success in high technology, and the important role that business culture plays in that process.

- The evolution and integration of a business strategy that exploits the many sources of competitive advantage available to the high-technology company.

Emphasis

Throughout, I emphasize the importance of the day-to-day operating tactics as well as the strategic decisions that ultimately determine the success of the technology-based business in seizing market opportunities, improving competitive posture, and developing and capitalizing on new technology; it is more the operating tactics than the broad corporate strategy that distinguish high-technology companies from other types of companies.

Organization

The book is organized around the interrelationships that both drive the complex, high-technology business enterprise and at the same time create competitive opportunities.

Chapter 1 emphasizes that, by their nature, technology-based corporations are complex systems. That complexity becomes particularly evident as one focuses on the need for interaction and coordination across the traditional functional areas of the business: engineering, marketing, and manufacturing.

Chapters 2 through 4 examine these interfunctional areas that the successful technical company must be particularly adept at coordinating. Chapter 2 examines the first important working relationship, that between marketing and engineering. Particular emphasis is here placed on defining new product opportunities and specifications, and also on marketing issues that arise due to technology being a key element in the overall marketing strategy. Chapter 3 examines the interdependence of marketing and production, while Chapter 4 investigates the traditional battleground between engineering and production.

This succession of discussions—engineering-marketing, marketing-production, and production-engineering—does not, however, quite do justice to the complexity of the technology-based organization. To round out that picture is the purpose of the remaining chapters. Chapter 5 explores the pervasive issue of quality, while Chapters 6 and 7 focus on financial returns, financial structure, and financing alternatives. Chapter 8 examines implications for the organizational structure as well as key general-management policies arising from the need to foster creativity and risk-taking within the company.

Finally, Chapter 9 presents an analytical framework for the key strategic questions that face top managers within technological business units. The emphasis in this capstone chapter is on developing and implementing an integrated strategy that both exploits the competitive leverage inherent in technology and that is at the same time both dynamic and evolutionary.

Acknowledgments

My indebtedness runs both deep and wide. My early mentors within high-technology companies should certainly recognize in this book the extent of their teaching and their influence. My colleagues both in the academic world and on various corporate boards of directors must also take credit for many of the ideas and admonitions contained in this book. Graduate students and participants in executive education programs who have not been content with superficial explanations have, by their questions and challenges, forced me to a more explicit level of thinking and explanation. Thanks to them all. And, finally, thanks to the editors at Lifetime Learning who encouraged me to write this book—and then challenged me to improve on my early efforts.

Henry E. Riggs

Chapter One

High-Technology Companies

This chapter covers the following topics:

- What are Technical Companies?
- What Do They Look Like?
- Managers in High-Technology Companies
- The Importance of Technology-Based Businesses
- Invention and Innovation
- Worldwide Competition Stimulates The Need for Better Managers
- Operating General Managers versus Strategic Planners
- New Technical Ventures
- The Complexity of High-Technology Companies
- The Objective: An Integrated Strategy

All individuals and enterprises use technology. Thus, in some sense, all companies are technical companies. Fast-food retailers use electronic cash registers and microwave ovens. Banks use automated and computer-based tellers. The wine industry—about as old an industry as exists—employs sophisticated technology to control its processes, to monitor quality, and to forecast yields, market demands, and inventory positions.

Some companies, however, create technological products, processes, or services. For these, technology is a key—and often *the* key—competitive tool. Managers in these companies face challenges and opportunities that derive specifically from this dependence upon technology. It is to these managers that this book is addressed.

1

◆ *What Are Technical Companies?*

The distinguishing feature of a technical company is that technology is a key element of its business strategy. Technology is certainly not the sole strategic element or even necessarily the dominant one; market, financial, and manufacturing considerations also drive the overall business strategy. But in a high-technology company, to formulate strategy without careful attention to the opportunities offered and threats posed by evolving technology would be analogous to assembling a symphony orchestra without a string section.

This definition suggests that a computer company or an integrated-circuit company is a technical company. An oil company is not because its strategy is typically geared to exploring and marketing and not to new product or process technology. A producer of lasers or laser-based systems is a technical company, but the supermarket that uses laser-scanning checkout systems is not. The laser manufacturer's business strategy will be a function of the forecasted evolution of laser and related optical technologies; the supermarket's business strategy need consider only peripherally, if at all, the changes in the technology of checkout stands, concentrating its attention instead on issues of product selection, promotion, store location, and customer demographics.

Where is your company on the "technologyness" spectrum?

Bioengineering or genetic engineering firms are clearly technical companies; apparel merchandising and frozen food manufacturing are clearly nontechnical. Other companies are not so easily categorized, and some that have not aggressively used technology in their business strategies might benefit from doing so—that is, by redefining their business along more technical lines. For example, U.S. auto manufacturers have not operated over the past several decades in accordance with my definition of technical companies. Their strategies have been dominated by marketing (and, arguably, regulatory) considerations, with neither product nor process (manufacturing) technology accorded much attention.

In reality, then, the dichotomy between technical and non-technical companies is not so clear. All companies are affected by technology—even the janitorial service firm and the fast-food purveyor—but some are much more affected than others. If all companies were arrayed on a spectrum of "technology-ness," our focus would be on one end of that spectrum: on those companies most dependent upon technology as an element of strategy and as an effective competitive tool. Such companies tend to be labeled "high-technology companies."

In this book, I am dealing with a particular set of technology-based operations: those in the private sector. The public sector of the economy presents many interesting and challenging technical management issues. Consider, for example, managers in the Department of Defense or Department of Energy in the federal government or managers of highway or other public works projects at the national, state, and local levels. My focus, however, is on the private sector: competitive industry.

My primary emphasis is not on those industries that are characterized by monopoly (for example, the regulated power utility or telephone companies), oligopoly (an industry dominated by a very few suppliers, such as the auto industry), or monopsony (an industry that has a single buyer, such as the national defense industry). Rather, my attention is on those industries characterized by multiple suppliers and many customers, where competition is intense (but by no means limited to price competition) and the barriers to entry are not insurmountable.

I raise the issue of barriers to competitive entry frequently. Entry barriers are often perceived to be more impenetrable than they are. A fundamental patent may be an effective barrier, but most patents simply require the aspiring competitor to find an alternative solution to a customer's need. Although capital requirements or dominant market positions are often mentioned as competitive barriers, recent successful start-ups of new companies in large mainframe computers, airlines, air freight, and communication equipment and services provide ample evidence that barriers to entry in those industries were not insurmountable.

More examples in the book are drawn from companies providing technical products than from those providing technical services, but the line between technical products and services is at best blurry. Many technical products require presale or after-sale service, and companies marketing such products would do well to consider themselves in both the product and the service business. Some technical services—for example, computer software or long-distance telecommunication—have many of the features of a product.

➤ *What Do They Look Like?*

What are some distinguishing features of high-technology companies? First, they tend to be well populated with engineers. Those on the cutting edge of technology are likely to have a fair sprinkling of research scientists as well. The research and engineering functions probably consume a substantial portion of the firm's resources. Engineers are also likely to be important members of the sales and marketing departments because an understanding of the company's technology base is a prerequisite to effective marketing. Engineers assume key roles in manufacturing because issues of producibility and reliability have important technology overtones. And, of course, top management positions typically go to those individuals who are, by formal or informal training, engineers.

High-technology companies are full of engineers.

A second distinguishing feature of high-technology companies is that product life cycles are likely to be short. The primary products or services offered today may well not have existed five years ago. And those that will comprise the bulk of the company's business in another five years may be today only a gleam in a development engineer's eye. In high-technology industries, competition is thus robust. For example, only about 25 percent of Hewlett-Packard's sales in 1981 were derived from products that had been offered for more than five years, and more than one-third came from products introduced in 1980 and 1981.[1] Changes in products, changes in technology, and changes in competitive positioning are ever-present facts of life in high-technology companies.

Change spells risk. Thus, a third distinguishing characteristic of high-technology business is riskiness. A company producing a single high-technology product sold to a single set of customers faces a precarious corporate existence, all the more so if the technology has little or no patent protection. However, the degree of risk should not be overstated. A broader based company, one that is a composite of a reasonable number of high-technology business units, may have a combined corporate risk in line with that of low-technology companies. Hewlett-Packard, Texas Instruments, Syntex, Itek, and many other large high-technology companies are diversified across a range of customer markets. Although particular product lines or divisions may be characterized as high risk, the corporations in the aggregate are not viewed (at least by investors) as unduly risky.

Competition is robust and business is risky in high-technology industries.

High risk and rapid change, in turn, spell instability. Instability spells opportunity to the optimist and danger to the pessimist. A fourth feature of high-technology companies is that they are more likely than low-technology companies to face rapid growth or rapid decline. Both eventualities pose particular challenges for management. So we will look at the financing, human, and facilities challenges brought on by rapid growth and the financial and emotional crises that frequently attend sudden retrenchment.

Instability presents both dangers and opportunities.

Dr. Simon Ramo, a successful founder and manager of a high-technology company, summarizes the distinguishing features of a technological company:

> [Every] corporation is an intricate system of people, machines, and facilities. It is a marvelous network of flows of information, money, material, and products. It has myriad connections with the outside world through which a pattern of . . . influences is communicated. If it is, moreover, a *technological* corporation, then it is even more marvelous and complex and also more dynamic and puzzling. Its laws and nature are even more intriguing and challenging to observe and understand. Now the very heart

of the corporation's activity—the development, production, and marketing of technological products—is in constant disturbance as a result of scientific discovery and technological advance.[2]

 # Managers in High-Technology Companies

Managers in high-technology companies are not a fundamentally different breed than their counterparts in low-technology companies. Their objectives, motivations, rewards, and responsibilities may be virtually identical. Many of their tasks will also be the same. But when technology is a key to business strategy, managers face other challenges and opportunities that require them to perform tasks and to delineate corporate practices that are fundamentally different from those in companies residing at the somewhat more peaceful end of the technology spectrum.

 # The Importance of Technology-Based Businesses

High-technology is critical for future prosperity.

Although I focus on a relatively small segment of all business enterprises, this segment—technology-based businesses—is critically important to most economically developed nations, and particularly to the United States. Consider for a moment those areas of the United States where high-technology companies have flourished: the so-called Silicon Valley in Northern California, Route 128 around Boston, Massachusetts, and

Table 1-1. Comparison of productivity and growth.

	High-Technology Companies	Low-Technology Companies
Annual increase in productivity	4.0%	2.0%
Compounded annual growth rate	6.7%	2.3%
Annual growth in employment	2.6%	0.3%

certain areas of the South. High technology has contributed greatly to the economic prosperity of these regions. Indeed, a study conducted at the Massachusetts Institute of Technology of a sample of low-technology and high-technology companies from 1950 through 1974 revealed significant differences in productivity and growth, as Table 1-1 shows.

Improvements in productivity and the creation of new jobs have been and continue to be important national priorities in the United States and in many other countries. A strong case can be made that technology-based businesses are the key to improving a developed country's economic health and competitive position in world markets.

Invention and Innovation

Our attention must not be limited to those large technical companies having the facilities and financial resources to undertake basic scientific research from which marketable new products may or may not be derived. Such basic research is immensely important, whether conducted in the private sector by such industrial operations as Bell Laboratories, DuPont, and IBM in this country, or Phillips, ICI, and Siemens in Europe, or by public (or quasi-public) institutions such as universities or government-sponsored research laboratories. The output of such basic research is invention: the conception of a brand new idea. But of parallel importance is innovation: the exploi-

Invention is the conception of a new idea.

tation or introduction of a new idea—that is, the commercialization of technology.

Smaller technical companies in the United States have an outstanding record of innovation. Indeed, several factors point to small companies being better able to innovate (exploit or introduce) than large companies. Niche markets that initially appear to large companies to be trivial in size may be appealing to small companies, and some of these apparently small markets—for example, personal computers and reprographic equipment—have exploded into major markets. Furthermore, the smaller company is probably better postured than the large company to perform the essential coupling between marketing and engineering that I discuss in Chapter 2.

Innovation is the commercialization of a new idea.

Thus, although invention often emanates from the laboratories of large companies, innovations with respect to that invention frequently occur in smaller technical companies. For example, solid-state electronic devices were not invented by a small company, but a series of small technical companies, some of which have grown to significant size, exploited the invention. Xerography was not invented by the small (and now large) company that exploited it, and minicomputers were brought to the market by small technical companies, not by the giant of the computer industry. Both inventions and innovations are critically important to the long-term competitive positioning of a country's technological industries in world markets.

Companies that are both technical and small create many new jobs.

Not only are high-technology companies better job creators than low-technology companies, but small companies are better job creators than large ones. More than 75 percent of new jobs created in the United States are created by companies of fewer than fifty employees. Another interesting statistic: The Fortune 1000 companies have in the aggregate over the past twenty years created no new jobs. Thus, companies that are both technical and small are of singular importance to a developed nation's economic health.

Our concern, then, must be with the management challenges in both large and small technical companies and in both inventive (those engaged in basic research) and innovative (those exploiting, or bringing to the market, new ideas conceptualized by others) technical companies.

Worldwide Competition Stimulates The Need for Better Managers

Until recently, the United States has been the world leader in technology. Today technology-based companies in Japan, Germany, and elsewhere—sometimes with the active support of their governments—threaten that leadership in many industries. The most highly developed economies of the world will increasingly depend on knowledge-based industries, with labor-intensive and raw material-intensive industries tending to shift to developing countries. Knowledge-based businesses are typically technology-based businesses. Therefore, the relative success of developed economies in world markets—and regional markets are increasingly giving way to world markets—will depend upon the skill and foresight with which these technology-based businesses are managed.

High-technology markets are worldwide.

Business management in the United States—and particularly the formulation of business strategy—has for many years emphasized marketing and financial considerations. Evidence of the overriding influence of marketing and finance in corporate strategy includes:

1. Proliferation of product lines offered in many industries, with little stimulation of primary demand for the industry's output
2. Corporate conglomeration, the combination of fundamentally dissimilar businesses in the often vain hope of gaining some financial advantage
3. Reliance on acquisition rather than internal product development to gain entry to new markets
4. Reduced investment in industrial research and development as a result of management's preoccupation with short-term financial results

By and large, these management policies have not resulted in long-term success for the companies employing them. Real growth (net of inflation) has often been minimal or negative for these companies and they have been decidedly unsuccessful in creating new jobs.

Managing high-technology companies demands new skills.

In contrast, Japanese managers—and the better managed companies in this country—seem able or willing to take a longer term and somewhat less financially dominated view of their businesses. It is now clear that Japan, often thought to be a technology follower rather than a leader, is spending a larger portion of its resources on technical development than is the United States and, not so incidentally, is training proportionately more engineers.

Evidence in the popular and management press suggests that managements in all developed economies are increasingly alert to the need to improve productivity, to innovate, and to seize opportunities offered by new technology. The objectives of engineering and manufacturing are no longer viewed as simply to support the business policy and strategy defined by the marketing and finance specialists. Increasingly, these functions are reclaiming their rightful role as key contributors to the business strategy.

The widespread discussion of the need to "reindustrialize" America recognizes the need to improve the technology base of U.S. business.[3] Professor Martin Starr of Columbia University states that the nation's industrial base

> is going from hell-in-a-handbasket to worse, and the economists don't have the slightest idea what to do about it. What the country needs is a trained cadre of managers who can go into factories, analyze what's wrong, and then take steps to correct it. Engineers can't do that because they don't have the management and finance background, and finance specialists can't do it because they lack the technical knowledge. Only by producing managers who can apply a broad knowledge of business and technical processes to manufacturing problems will U.S. industry regain its competitive edge.[4]

Engineers must be trained to manage.

The objective of this book is to assist in that process of training engineers to manage. The technical and engineering view must be represented in the councils of top management, and engineers must gain a general management perspective if they are to be effective in those councils.

Operating General Managers versus Strategic Planners

The perspective in this book is that of the operating general manager—the manager who must integrate and orchestrate business functions. This individual may not be the corporate chief executive; indeed, in large companies, he or she is almost surely not the chief executive.

The job of the chief executive officer (CEO) of a Fortune 500 industrial company is to deploy corporate resources across a broad range of present and potential business units, with particular attention to financial objectives and constraints and to changes in the external environment that create problems and opportunities. That is, the large company CEO is a strategic planner, not an operations manager. Acquisitions and divestitures, balancing the portfolio of business units, and corporate reorganizations to combine or redefine business units occupy the CEO's attention. Seldom is he or she concerned with such issues as (1) coordinating within a particular unit the market opportunities the marketing department perceives with the technical opportunities the development group perceives, (2) the need to strike a balance between responding quickly to customer demands and reducing product manufacturing costs, or (3) retaining manufacturing flexibility while capitalizing on new technology as it becomes available. These issues are left to the business unit manager, typically a division manager.

It is the business unit manager—in a large company, perhaps the head of a division and, in a small technical company, probably the president—who carries out those tasks peculiar to the technology-based businesses. Interestingly, even the product manager of a line of technical products—a middle management position—may carry many of the responsibilities of a business unit general manager. Whatever his or her title, the business unit manager must develop strategy and tactics appropriate to the particular set of products and markets the business serves. That strategy and those tactics occupy our attention.

New Technical Ventures

Many of the most successful technical ventures in recent years have been companies created by entrepreneurs to capitalize on some particular piece of technology or on some market niche not served by the giant companies. Evidence abounds that entrepreneurial technical ventures are particularly able to exploit new market and technological opportunities. Many new industries, including minicomputers, process control, laser instruments, and bioengineering, have really been created, and are now dominated, by companies founded by entrepreneurs. The venture capital industry, an industry more highly developed in the United States than in any other country in the world, has focused primarily (but by no means exclusively) on investments in new and emerging high-technology companies.

I do not address the entrepreneurial function per se in this book. I leave to others the discussion of the initial tasks of assembling the team, obtaining the financing, clearing the various legal hurdles, and providing the spark, enthusiasm, leadership, and hard work that brings together all the pieces of the new enterprise.[5]

Present and potential entrepreneurs should bear in mind, however, that, following the initial rush and excitement associated with the founding of a new business, the technical entrepreneur becomes the general manager of a technology-based business. Some foresight as to the problems and challenges inherent in managing a technical business will serve the entrepreneur well as he or she considers the mix of personnel to be recruited, the markets to be served, and the money to be raised for the new venture. The management challenges

Entrepreneurs become general managers of their high-technology businesses.

evolving as the new technical company grows from a single-product, single-market company to one offering a range of products to several markets—and from one dominated by the entrepreneur to one that is more classically well managed—commands a good deal of attention in upcoming chapters.

The Complexity of High-Technology Companies

The very nature of high-technology companies results in their being complex organizations. They must be able to market effectively; yet what they market must be a function both of what the customers need (or can be projected to need) and of what available (or emerging) technology is able to provide. Technology-based companies must be effective innovators and developers of products and services, but what they develop must be producible at an acceptable cost and must conform to specifications that are realistic to the development engineers and useful to customers. These companies must be efficient producers, but efficiency must be viewed in light of the customer's demand for quality and delivery, the justifiable investment in development engineering and tooling, and the probability of changes in technology in the months and years ahead. Financial returns must be consistent with the risks in the business and the requirements for investment in new product development and working capital.

High-technology companies are complex.

These complexities seem to manifest themselves particularly at the boundaries between the traditional functions of industry: engineering development, production, and marketing. The lives of both the general managers and the functional managers would be greatly simplified if (1) the development engineers could be allowed to develop products without concern for the market or the producibility of the products, (2) the planning and execution of the production activity could

Complexity places special demands on managers.

be independent of engineering and marketing, and (3) marketing could specify, promote, and sell products without ever communicating with engineering or production people. Some technical companies seem to operate in such a manner, with very little cross-departmental communication or cooperation. But they are rarely successful for long, particularly in an environment where rapid technology changes induce changes in customer demands and in competitive product offerings.

The management complexities in technical businesses, although a distinguishing feature, need not be viewed as a disadvantage. Indeed, they represent a kind of beauty, just as art or music or nature may be particularly intriguing because of its complexity. Moreover, this very complexity offers an opportunity for the successfully managed technical company to distinguish itself from its competitors. Successful management of these boundary complexities is a key competitive weapon.

The complexities are not limited to those that exist at the interface between just two functions. The market segment targeted for the technical product affects the volume to be produced, which influences the appropriate production process and thereby the optimal engineering design sophistication. In turn, the product and process design and required investment dictate cost and thereby the distribution methods as well as the market segments that can or should be tapped. Top management's key task is to orchestrate among the conflicting demands of *all* business functions—engineering, production, marketing, and finance—and among all constituencies—customers, employees, stockholders, suppliers, and communities.

This orchestration must occur in an environment in which change is the dominant theme. The fact that high-technology companies are characterized by rapid growth, turbulence in both markets and technology, and the need to stimulate innovation and creativity places extra demands on general management to infuse the organization with an effective set of policies, procedures, objectives, and philosophies, the combined set of which is now often referred to as *organizational culture.*

➤ *The Objective: An Integrated Strategy*

The technologist who assumes general management responsibility carries also the responsibility for strategy. Indeed, the business unit strategy is defined by the way the general manager guides the organization in dealing with the inherent complexity of the business.

From time to time, the general manager needs to stand back from the day-to-day fray and ponder some key questions. Is the business' strategy well integrated? Does the strategy pursued in technology mesh well with both the marketing and the manufacturing strategies, and is it consistent with the financial constraints under which both the company and its competition operate? To what extent should the company be market centered rather than technology centered? What are the trade-offs between further concentration in particular markets and technologies and diversification into new products, markets, or technologies? I address these questions throughout the remainder of the book. The final chapter reemphasizes the necessity of formulating an *integrated* strategy within the high-technology company.

Complexity demands that business strategy be integrated.

Highlights

- Technology is a key to the business strategy of a technical company.

- Technology is a key competitive weapon.

- High-technology companies are vital to economically developed nations.

- High-technology companies are distinguished by the presence of many engineers, short product life cycles, high risk, and rapid change, all of which create instability.

- High-technology companies are more complex than low-technology ones.

- Both invention and innovation are critical in high-technology industries.

- Complexity demands that general managers stress the integration of the operating functions of the high-technology business.

Notes

1. Hewlett-Packard Company, *1981 Annual Report* (Palo Alto, California).

2. Simon Ramo, *The Management of Innovative Technological Corporations* (New York: Wiley–Interscience, 1980), p. 1.

3. "The Reindustrialization of America," *Business Week* (June 30, 1980).

4. *Wall Street Journal* (January 26, 1981).

5. See, for example, Patrick Liles, *New Business Ventures and the Entrepreneur* (Homewood, Ill.: Irwin, 1974) or Karl Vesper, *New Venture Strategies* (Englewood Cliffs, N.J.: Prentice-Hall, 1980).

Chapter Two

Coordinating Marketing and Engineering

This chapter covers the following topics:

- The Critical Decision: Choosing Product-Market Segments
- Forecasting: A Joint Responsibility of Engineering and Marketing
- Customers: An Often Overlooked Source of Innovation
- Patents: An Unreliable Barrier
- Categorizing Products by Use and Degree of Standardization
- Application Engineers: At a Critical Boundary
- Selling Technical Products: By and to Engineers

Tension at the boundary between the marketing and the engineering functions in technology-based business is both obvious and widespread. But this tension can be productive, and management must make it so by paying close attention to the coordination and cooperation between these two key functions.

In this chapter, we consider choosing product-market segments and the shared responsibility of engineering and marketing in making these choices; techniques for perceptive forecasting of both technology and market evolution to assist in those product-market choices; an often overlooked source of innovations—the customers—and ways to organize to capitalize on this source; a system of product categorization that facilitates further analysis; the application engineering function, which assumes great importance because most technical products consist of a mix of hardware and software (or service); and some key marketing challenges that attend technical products, particularly in regard to technical communication and signaling between supplier and customer, and the marketing of radically new products.

The Critical Decision: Choosing Product-Market Segments

Selecting the product-market segments to pursue is arguably the most critical management decision for technical companies. Management's challenge is to be certain that these product-market choices do not suffer from either marketing or engineering myopia. Examples of product, and even corporate, disasters abound; their cause can typically be traced to inadequate attention to either the needs of the market or the capability and state of the technology. A major office equipment manufacturer introduced an office copying machine incorporating technology far superior to the competition, technology that resulted in outstanding performance specifications. Unfortunately, potential customers were indifferent to the particular specifications offered and rejected the product entirely at the price that the manufacturer had to charge.

Avoid the disasters that can result from marketing or engineering "myopia."

Hovercrafts have been an engineering miracle and a commercial disaster, as has the supersonic transport airplane. Polaroid's instant home movie system is a technical marvel, but it was finally withdrawn from the market for lack of customer acceptance. The files of venture capital investment firms are littered with tales of companies started by brilliant technologists who have designed products with no attention to market needs.

Disasters are not caused solely by engineering mypoia. Failures arising from product-market selections made solely with a view to the market also abound. These disasters tend to be somewhat less visible, although not necessarily less costly, because the product, although designed, is not successfully produced. If it is produced, it is not successfully sold. In the early days of computers and more recently in process-control equipment, automated merchandising systems, and robotic systems, suppliers have made promises of performance specifications that were simply unachievable with available technology. The persuasive salesperson or sales manager—and the capable ones are all persuasive—can convince both the customer and his or her own top management of the need for a particularly demanding set of specifications. If these specifications cannot be achieved, or can be achieved only at exorbitant cost, the company will suffer both unrecovered costs and, sometimes more important, customer ill will.

A key challenge, then, in all technology-based companies is to couple engineering and marketing considerations in choosing which product-market segments (or niches) to pursue. But where do new product ideas and the perception of new market niches arise? Can they be forecast? What can management do to stimulate these new ideas and how should they be investigated? How can management ensure that both the technical and marketing viewpoints are adequately represented when making these key decisions? These questions occupy our attention for much of this chapter.

The choice demands a coupling of marketing and engineering views.

The Respective Roles of Engineering and Research

First, what is the technical source of new or improved products that are the lifeblood of high-technology companies? Is it the research department or the engineering department? In Chapter 1, I distinguished between invention and innovation, and this distinction bears repeating here.

Invention—the conception of a brand new idea—is the product of research. The most fundamental research is undertaken with the realization that a broad range of outcomes is possible, with some having little or no commercial or economic significance. Because one cannot know the dimensions or characteristics of the new idea before it is conceived, fundamental research, although important, can be guided in only the most general of ways by considerations of market needs.

Innovation—the exploitation or introduction of a new idea—is the product of engineering. The introduction of a known idea should be guided by an understanding of, and appreciation for, the needs of the market.

As illustrated in Figure 2-1, invention and innovation—and therefore research and engineering—are at two ends of a technology-source continuum. The most basic research, the

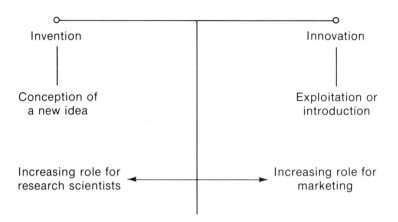

Figure 2-1. Technology-source continuum

Invention and innovation demand different technical and managerial skills.

practical consequences of which are unknown in advance, lies at one end of the continuum. At the other end is engineering of the most applied kind, for example, simple modification of an existing product. The boundary between invention and innovation, and thus between research and engineering, is both unclear and irrelevant. What is critical is to recognize that *both* are important; that the farther one is operating toward the applied engineering end of the continuum, the *greater* the need for close coupling between marketing and the technical function; and that the management style and process will be different, depending upon where the company lies along the continuum.

Ansoff and Stewart address this last point in a classic article.[1] They contrast the characteristics of what they call R-intensive companies with those of D-intensive companies (see Table 2-1). Several contrasts point to the much greater need for market coupling in the D-intensive business.

Much technical work that carries the label of research is of a very applied nature and is more correctly designated engineering or development. Many companies whose annual reports glorify their R&D (research and development) efforts are in fact engaged almost solely in "D." Although the decision as to the appropriate mix of research and development is an important one strategically, the operating manager's attention needs

Table 2-1. Characteristics of the engineering function in R-intensive and D-intensive companies

Research-intensive Organizations	Development-Intensive Organizations
Work with indefinite design specifications	Well-defined design specifications
"Broadcast" objectives and market data among technical people, rather than channel specific kinds of information to individuals	Highly directive supervision
	Sequential arrangement of tasks
Nondirective in work assignments	Vulnerability to disruption from changes in objectives or specifications
Maintain a continuing project evaluation and selection process	
Stress the recognition (interpretation) of significant results	
Value creativity over efficiency	

Source: Adapted from H. Igor Ansoff and John M. Stewart, "Strategies for a Technology-Based Business," *Harvard Business Review* (November–December 1967), pp. 71–83.

to be on the coordination between engineering (as opposed to research) and marketing.

Do not infer that research is unimportant, unproductive, or unwise for the technically based company. Although breakthroughs in science are seldom in themselves the source of new technical products, invention may yield technology that in turn will (with some time delay) result in new product or process opportunities. The point is simply this: Innovation—exploiting known technology for new or existing market needs—is the joint responsibility of engineering and marketing.

Innovation's Many Parents

Product innovations are popularly thought of as resulting from carefully planned and properly budgeted engineering projects that build on technology that is proprietary to the company. This conception suffers on at least three accounts.

Don't rely solely on engineering projects for useful innovations.

First, product-market innovation (but probably not invention) can occur as well within the marketing staff as within the technical staff. No fundamental technical breakthroughs are required. Exploitation of the technology is as likely, or more likely, to be the inspiration of a person devoted to the marketplace as of a person devoted to the laboratory.

Second, some of the most exciting new ideas—whether at the invention or innovation end of our continuum—come forth in an unplanned, unbudgeted, and unorthodox manner. "Bootleg" or "pet" projects in engineering (unauthorized and unbudgeted by management but frequently tolerated or even subtly encouraged by organizational norms and operating practices) may provide the initial impetus toward a new product or new technological strength. This first step may, over time, change the fundamental strategy of the business unit.

Third, the role of *reverse engineering*—an innovative process held in disrepute by some—is often important. Reverse engineering is the process of working backwards from a competitor's product to incorporate its technical advances in one's own product. When reverse engineering involves espionage or the stealing of secrets, it is at least unethical and probably illegal. When ideas are derived from competitors' products available in the market, reverse engineering is simply one of the mechanisms of technology diffusion.

Managers should be aware and accepting of innovations that arise from unconventional sources via unconventional means.

Types of Risks

Development projects are indisputably risky; perhaps only 20 percent of them result in commercially successful products or processes. When the project involves state-of-the-art technology, the risks of failure are even higher.

Overall risk is the product of three factors.

To a great extent, however, risks associated with new products are more commercial than technical. The total risk associated with new product development and introduction can be thought of as the product of three separate risk factors:

1. The probability of technical success—that is, the risk that the technical problems can be solved in a manner that will result in a product or process that meets (or closely approaches) the desired specifications
2. The probability of commercialization, given technical success—that is, the risk that the specifications sought will not prove adequate, that costs to produce are excessive, or that management will decide for any of a number of reasons (including cross-elasticities, competitive positioning, and timing) not to introduce the product
3. The probability of economic success, given commercialization—that is, the risk that the new product will not prove sufficiently salable to yield a rate of return on development (both engineering and marketing development) at or above the firm's hurdle rate for new investments[2]

In the light of the discouraging track record of new product introductions, the challenges to engineering and marketing executives in high-technology companies are to do the following:

1. Forecast both technology and markets so that the right product-market segments are pursued
2. Monitor throughout the development project both the progress of the project (and prospects for meeting the target specifications on time and on budget) and the evolution of the market so that, as the probabilities of ultimate success of the new product shift, appropriate corrective action can be taken to redirect, halt, or accelerate the development

∴ Although all functional areas of the business are involved in these difficult risk assessments, the primary task falls to engineering and marketing. Coupling these two functions is as vital in assessing risks and opportunities of new product development as in choosing product-market segments.

Forecasting: A Joint Responsibility of Engineering and Marketing

A key challenge facing all top managers is choosing which product and market segments to pursue. The challenge is heightened in a high-technology company because often, when the choice is made, neither the product nor the market exists. The product does not exist because no one has yet exploited available or emerging technology to produce it. The market cannot exist if the potential customers are unaware of the product's existence. Somehow the ability to design and produce the product must be forecast, and the market potential must be assessed, all with very little data. This absence of data is typically more acute the newer the technology in which the company is engaged.

Forecasting must occur before either products or markets exist.

Forecasting the home or personal computer market is an interesting case in point. Until home computers were available, users and various software suppliers were uncertain about useful tasks they might perform, how many entertaining games would be forthcoming, or how appealing the games might prove to be. Any forecast of the home computer market would have been incomplete without a forecast of the technology upon which the computer depends. The opportunities in this market were going to be profoundly influenced by the evolving technology in software, in input-output devices, and in the microcircuit industry, particularly in large scale integration and memory technology. Moreover, a technology forecast that ignored the dynamic growth and evolution of the market was likely to be just about as inaccurate as a market forecast that assumed no changes in technology.

The truly exciting new product-market opportunities arise when management forecasts an intersection of the potential of new technology with the needs of a set of customers. Making such a forecast requires broad and comprehensive knowledge of both the relevant markets and the relevant technologies. It also requires hard work, an ability to synthesize information from various sources, and a willingness to take risks. The most successful technically based companies do this forecasting well, thereby coupling technical knowledge and prophecy with market knowledge and forecasts.

Opportunity lies at the intersection of new technology and evolving customer needs.

As I discuss specific forecasting techniques, keep two caveats in mind. First, although I discuss separately some techniques of technology forecasting and market forecasting, general managers face the important job of integrating the two forecasts and seizing the new opportunities thereby brought to light.

Second, we must be realistic about the accuracy, and therefore the usefulness, of long-term forecasts. One authority on the subject says,

> The track record of many long-term forecasts has been pretty dismal. . . . Although there is little hope of increased accuracy in long-term forecasts, it seems sensible to
>
> 1. Avoid trusting any particular forecasting method, regardless of how convincing it or its results may appear to be.
> 2. Think seriously about the true significance of recent events; seemingly important events can ultimately prove trivial, and vice versa. . . . After all, it should be remembered that the future, even the long term, begins with the present.
> 3. Look for reasons why your forecast is likely to be wrong.
> 4. Look for the implications of their being wrong.[3]

Technology Forecasting

A primary responsibility of both top management and senior engineering management in a technology-based company is forecasting the changes in technology that do or may affect the business. A top executive who fails (or refuses) to immerse himself or herself in the relevant technology is unlikely to make wise product-market choices. Technological threats will be ignored until too late, and opportunities created by the evolution of technology will be overlooked.

Unfortunately, in many companies where technology could be an important element in strategy, the top executives shy away from involvement with technological issues and decisions. Some observers have attributed this reluctance to a preoccupation with financial parameters for decision making. Another reason may simply be a fear of the unknown: Top managers who are not trained (formally or by experience and reading) in technology will avoid exposing their ignorance; they will delegate technical decisions to others. Where technology is fundamental to the business strategy, complete delegation is clearly inappropriate.

Technology forecasting can be neither delegated nor subcontracted.

The forecasting process is at best imperfect. Clairvoyance is desirable, but seldom available. Although hiring experts for counsel may be helpful, hiring them to make decisions seldom works. Experts will insist on accompanying their recommendations with data and evidence. Data are scarce, and by the time evidence becomes wholly convincing, the time for action to seize opportunities or avoid dangers is probably past. Technology forecasting is an art, not a science.

The need for sound forecasts of technological changes is indisputable. When those forecasts are derived from systematic reasoning, they improve the resource allocation decisions in technical companies—what development paths to pursue, what process technologies to invest in, and what marketing channels and capabilities to develop. When these decisions are made (as they too frequently are) in the absence of systematic evaluation of technological changes, the implicit assumptions may represent little more than speculation, or even prophecy, about the future state of technology.

Where does one look for new technology? Certainly in one's own company—in the progress and output of specific development projects, but often, more important, in the ruminations of the most creative engineers on the staff. One looks outside as well: at suppliers, at customers, at competitors (to the extent one can obtain knowledge short of industrial espionage), at companies in allied industries, and at universities.

Scan broadly for new and changing technology.

There is no substitute for extensive and broad reading of both the technical and the business press.

Monitoring the technology horizon is a key task for executives in technology-based companies. This monitoring must be undertaken with open and receptive minds and a company attitude that accepts new ideas, whatever their source, and does not dismiss as unworthy those ideas that were "not invented here." Japanese companies are particularly good at technology monitoring. The larger companies devote substantial time and effort to staying abreast of new developments in their own industries as well as among suppliers and customers, both in Japan and around the world. Interpretation is the key. What are the implications of the reported or observed technological breakthrough or of the continuing evolution of an existing technology? For example, what are the implications to the petrochemical industry of the genetic engineering breakthroughs occurring in the pharmaceutical industry or the probable impact on the medical electronics industry of improvements in battery technology?

Importance of Systematic Reviews.   Technology forecasting should not be left to chance—an activity to be engaged in while commuting or showering. High-technology companies should undertake a systematic review, at least annually, of technological forces and trends both within and outside the industry that may present opportunities or threats to the company. Top managers, senior marketing managers, research and engineering managers, and chief scientists should participate in this review. Outside authorities (consultants, academics, or technically sophisticated members of the company's board of directors, for example) may be useful stimulants in the review process. Their questions, viewpoints, and diverse backgrounds can offset some myopia that almost inevitably becomes built into a company. The leaders of this review process need to be well read and some of the systematic forecasting techniques I discuss later in this chapter should be employed in advance or as part of this review.

The review process demands the participation of many.

Forces of Technological Change.   Two distinct frames of reference are useful in thinking about technological forecasting: technology-push and demand-pull. That is, some technical improvements and breakthroughs will occur—and can be

Technological change is both pushed by the technology and pulled by market demand.

forecast to occur—because of a continuation of the evolution of a particular stream of technology. Others will occur because the demand is evident for devices or products with certain performance specifications that have not yet been achieved but may be realized with additional development effort.

We know that computers will get faster and that programming languages will get both more efficient and easier to use. The timing, magnitude, and precise form of these improvements may be unclear, but present technology is pushing in their direction. Very large scale integrated (VLSI) circuits could be forecast well before they were realized because a forecaster could observe the trends in technology that were pushing in the direction of such devices. Some astute observers of technological trends today forecast a continuation over at least the next decade of the 30 percent annual decrease in cost per function of integrated-circuit (IC) devices—a technological forecast having enormous implications.

Demand-pull characterizes the technology environment for research in photovoltaic solar cells. The demand for cells of various efficiencies (electrical output as a function of cost) can be plotted. Expensive cells are now being used in spacecraft and certain other applications. As efficiencies improve, more active solar cells will be used, first in remote locations, where alternative sources of electric power are expensive, and then increasingly in more mundane applications. Both technologists and investors are aware of these opportunities. Demand will pull technological resources into the development of solar cells and the result will be major breakthroughs—and probably many false starts, as well.

More fuel efficient internal-combustion engines are subject to a similar demand-pull phenomenon. The need is evident and a forecaster can be relatively certain that technological resources will be expended to fill that need. Outside the field of energy, one might reasonably forecast that technology change in the residential building industry will sharply accelerate. As various resource constraints conflict with certain demographic realities, change will occur in spite of labor union and government restrictions that have stalled technological evolution in this industry in recent decades. The pent-up demand for housing will force the technology changes.

Systematic Forecasting Procedures.   Without engaging in a detailed discussion of the many techniques of technology forecasting, I can summarize several recognized and systematic techniques. Some of these are very qualitative and some seek to be relatively quantitative. Generally, the methods are best used in combination; the forecaster will not want to rely solely on one procedure. When several procedures point to the same forecast of technology, one can have increased confidence in the prediction. We should heed the advice of the technological forecaster who says, "To be useful, technological forecasts do not necessarily need to predict the precise form technology will take in a given application at some specific future date. Like any other forecasts, their purpose is simply to help evaluate the probabilities and significance of various possible future developments so that managers can make better decisions."[4] Even knowing the direction and trend of future technology developments may be of substantial value to management.

Use a combination of techniques for technology forecasting.

All techniques require a careful analysis of past experience and the best efforts of competent, insightful, and imaginative people. The following nine techniques are systematic procedures for thinking about the past and the future.[5]

1. *Contextual mapping* describes in a systematic way the evolution over time of the primary and associated technologies in a field. It permits a field to be surveyed with the hope of discerning trends in developments, as well as barriers to further development. The technique has been successfully used in the chemical and defense systems industries.

Use these nine techniques for analyzing the past to forecast the future.

2. *Gap analysis* involves a systematic review of a range of products, processes, or technologies to discern gaps in the range or series that is likely ultimately to be filled. The method appears to be particularly applicable to the observation of nature; it has permitted the forecasting of the discovery of new entries in the periodic table.

3. Another intuitive procedure is referred to as *forecasting by analogy:* finding an analogy between the technology to be forecast and some historical event or process. If the analogy is valid—a significant "if"—the former event can be used to predict the future development of some area of technology.

4. *Scenario writing* is "a method of harnessing a range of forecasting techniques to a study of the total environment as

it affects the attainment of corporate objectives."[6] The method has attained greater popularity in industry since the energy crisis of the 1970s as companies have attempted to assess the impact of wildly different forecasts of energy price and availability.

5. *Forecasting specific attributes or parameters* is a somewhat more quantitative approach to technological forecasting. It involves isolating one or a few key performance factors— that is, attributes or parameters—of a product (or family of products), such as computer executions per second or engine efficiency in miles per gallon, and forecasting in numerical terms the values of those attributes or parameters at some date in the future.

6. *Relevance trees,* or *technical planning flow charts* or *network analyses,* permit the forecaster to assess interdependent technologies and tasks with a view to isolating the critical developments that must be achieved to realize certain product or process objectives.

7. *Time-series forecasts* are based upon an extrapolation of historical quantitative data of an appropriate parameter. These projections may be simple or reasonably complex and must take into account natural limits (for example, 100 percent efficiency of a device), substitution effects, and other occurrences that may accelerate or retard the rate of change in the technology.

8. The *Delphi procedure* represents more an additional refinement than a fundamentally different method of technological forecasting. Delphi is a "set of procedures for eliciting and refining the opinions of a group" of experts.[7] Individual opinions are collected anonymously and fed back to the group of experts in summarized form. The procedure is typically repeated several times, and the final outcome is not consensus or agreement but rather the development of a statistical index of the collective opinions of the experts.

9. Other collective approaches to forecasting include *brainstorming* and similar creativity spurring techniques. These qualitative techniques have in recent years taken on more scientific sounding names, but any such technique "is conducted by a group of people who attempt to solve a specific problem by collecting all the ideas spontaneously contrib-

uted. The method depends on the freedom of thought . . . permitted and indeed encouraged; in the avoidance of criticism and of any premature evaluation of suggestions; and on the synergy arising from the interaction of the thought-provoking ideas that emerge."[8] Experience with brainstorming varies widely, and satisfaction with the technique seems to turn on the leader's capability, the panelists' enthusiasm and creativity, and whether productive new ideas in fact emerge.

This list is not exhaustive, and some of these techniques, or variations upon them, are referred to by a variety of names. Furthermore, the best of forecasts will have limited accuracy. James Brian Quinn enumerates four primary shortcomings of forecasts.[9]

Even the very best forecast has limited accuracy.

1. Unpredictable interactions of technologies
2. Unprecedented demands—computers, energy-efficient equipment, and office photocopying are reasonable examples
3. Major discoveries—lasers and recombinant DNA are examples—but Quinn further argues that "such breakthroughs do not often simply burst forth on the world; rather, they frequently result from long streams of work" spread out over a decade or more[10]
4. Inadequate source data on historical trends in technological development and on current and prospective deployment of technical resources

The record of business in forecasting very fundamental changes in technology, or in the environment that affects the development of technology, has not been very impressive. Interestingly, a widely read and respected report by two futurists, published in 1967, failed to forecast the energy crises that erupted a few years later; indeed, neither energy nor oil appears in the subject index of the book.[11] Changes in the energy picture have turned out to be a driving force in technology change over the years since 1973 and have created untold market opportunities as well as product threats.

An Example of Successful Forecasting. Measurex Corporation at its formation in 1968 offers an interesting example of successful technology forecasting. The company's founder forecast that the confluence of several technology trends would create an opportunity to provide turnkey, digital-computer-based process-control systems for the control of the paper-making process. First, price and performance of minicomputers would permit the dedication of a single computer to a single paper machine. Time-sharing on a large, mainframe computer would become progressively less appealing as the price-performance ratio of minicomputers continued to improve. Second, evolution of sensor technology would permit the physical properties of the paper to be sensed as it was being manufactured, with the information transmitted in digital form to the computer. Third, software technology would permit standardization of computer programs across many customer installations, with resulting economies.

As the system design evolved, Measurex had to decide whether to use core or rotating memory. At the time the decision was made, rotating memory was considerably cheaper, but also slower. Measurex elected to use core memory, based upon the forecasts that (1) core memory prices would continue to decline, (2) these memories would ultimately be replaced by integrated-circuit memories, and (3) the near-term cost penalty of core memories could be tolerated during the early years of the product when premium selling prices could be charged.

This set of decisions, based upon insightful forecasts of technologies, permitted Measurex to enter the market in 1970 with a standard, turnkey system that provided improved performance over existing analog systems and was substantially more cost-effective than systems assembled by paper manufacturers from components supplied by various vendors, including IBM. Of course, the evolution of technology did not cease when Measurex entered the market. The company has had to be continuously alert to opportunities for and threats to its product lines from ongoing evolution of the minicomputer, the advent of the microcomputer, advances in input-output devices and operator displays, and the introduction of large scale (and very large scale) integrated circuits.

Market Forecasting

Market forecasting for new technical products is a very different proposition than forecasting the sale of, for example, a new consumer product that is entering a defined market against known competitors.

Forecasting Customer Needs.   The newer the technology, the more difficult the forecasting task. New technology creates customer needs, or at least brings forth latent ones. The world existed without xerographic reproduction equipment; customers, and even the primary producer of such equipment, were only imperfectly aware of the role that copying machines would come to play in offices around the world. Original projections of demand for xerographic reproduction devices were ridiculously low—ridiculous in light of future events, but not so ridiculous in view of actual office practices at the time.

> *The newer a product's technology, the harder it is to forecast its sales.*

Thus, the forecaster of demand for new technological products must assess the impact that the availability of the product, with its inherently different—not just improved— capabilities will have on customer behavior. One can ask customers, but they typically will not know. Most customers are looking for incremental improvements in present methods and have not considered radical alterations in behavior that might be possible or desirable with new technology. Prior to the introduction and acceptance of word processing equipment, office managers were looking for improved typewriters—faster, with facility for easier corrections—and better filing systems. They were not looking for a fundamentally different way in which to organize and conduct the business of their offices. Yet to realize the full potential of word processing equipment requires just such reorganization.

> *Forecasts of products requiring changes in users' behavior are particularly challenging.*

In the world of microcomputers, it remains unclear just how radically our lives will be changed by the incorporation of computation capabilities in present or new products. How many so-called smart products will be forthcoming and how extensively and quickly will they be accepted, probably first in industry and then ultimately in the home? How about videotape recorders? Declining prices and increasing customer acceptance may change TV viewing habits. If they do, how will networks and independent producers respond to these changes? Will these changes, in turn, accelerate or inhibit the growth in demand for the videotape recorders?

Created Markets.   Markets can be created. By definition, no market exists for the products of a new industry prior to that industry's emergence. Missionary zeal on the part of leading and early entrants in an industry can accelerate its emergence and growth. Dr. Edwin H. Land exhibited just such missionary zeal in promoting instant photography (the Polaroid process). Without Dr. Land, instant photography would have been years later in reaching the consumer, and sales would probably have been limited to the affluent for a much longer period. Although large companies would appear to have the advantage of abundant resources to undertake these missionary roles (for example, IBM in many computer applications), independent entrepreneurs have, to a remarkable extent in this country, been responsible for providing the drive, leadership, and enthusiasm necessary for creating entirely new industries.

Creating new markets offers the greatest risks and opportunities.

Examples can be cited where original market forecasts were woefully conservative: xerographic reproduction equipment and home computers, both mentioned earlier, but also video games, certain new drugs, handheld calculators, and even fast-food hamburgers (where the technology incorporated into the fast-food delivery systems is not trivial). But the potential for exciting new technology can also be wildly overestimated: hovercraft (or ground-effect machines), digital watches, video disk playback units, and three-dimensional photography. Timing may be off: The early forecasts for the use of programmable robots has not yet been realized, and the early forecasts of the demand for fiberoptics and for the photographic technique called holography were at least premature. Only the future will reveal whether the expectations of some people—particularly investors—for the future of genetic engineering will be realized and, if so, when.

Forecasting Competitors' Actions. Forecasting the total market is only part of the challenge. Competitors will enter the market. Who will they be and what will be the relative strengths and resultant market shares of all competing firms? Will the market appear so large, or prove to be so large, that many major companies will enter, as has happened in the personal computer business and the computerized axial tomography (CAT) scanner business? When a potential entrant initially forecasts the market, the number and strength of future competitors is typically unknown. Although all market forecasts require an element of war-gaming (figuring out what the competition will do in response to one's actions), when both the market and the technology are new, the uncertainties can be extreme. What is going on in those development labs and when will it be completed? Those are the tough questions. (The importance of these questions frequently demands that a high degree of security surround new product development.)

Technical and marketing innovations don't stop when your product enters the market.

Conventional, statistical market forecasting techniques are less applicable in high-technology markets than they are in the forecast of demand for nontechnology products in existing markets. Demographic and other market segmentation data are useful, but deducing just which segments are relevant and what will be, for example, the price elasticities within those segments requires both a forecast of the technical specifications of the product or service to be offered and the specifications of competing products or services.

The possible appearance of substitute products or services also must be forecast. A major purpose of technological forecasting is to attempt to foresee the emergence of these substitutes. However, substitutes need not incorporate new technology. They may arise from a competitor's effective marketing, including aggressive pricing, or from an imaginative customer who institutes the substitution.

Forecasting and Decisions over a Product's Life Cycle. Market forecasting becomes progressively easier and more accurate as the product progresses through initial testing, rapid growth, maturity, and decline. Market forecasting continues to be

The purposes of forecasting change over the product life cycle.

Table 2-2. Operating decisions made over a product's life cycle

Stage of Product Life Cycle	Decisions
Preproduct	Allocation of development funds Acquisition of technology Personnel needs Distribution system needs
Product development	Product design Amount of development effort Supporting functional strategies
Market testing & early introduction	Optimum capacity Distribution strategies Pricing strategies
Rapid growth	Production planning Facilities expansion Changes in process technology Changes in marketing strategy
Maturity	Production planning and inventories Special promotion and pricing
Decline	Transfer of facilities Inventory control of obsolescing parts and spares

Source: Adapted from S. K. Mullick, G. S. Anderson, R. E. Leach, and W. C. Smith, "Life-Cycle Forecasting," in Spyros Makridakis and Steven C. Wheelwright, eds., *The Handbook of Forecasting* (New York: Wiley-Interscience, 1982), p. 275.

important in these later stages; so I return to the subject in the next and subsequent chapters. Table 2-2 outlines, for each of the product life cycle stages, the operating decisions that rely heavily on accurate market forecasting.

∴ Forecasting the demand for a partially defined product in an as yet undefined market is a dicey affair, with high business risk to the company and high personal risk to the forecasters. I return frequently to the issue of risk taking in high-technology firms. The prevailing atmosphere and management norms in the company must tolerate risk taking if new technological products are to be forthcoming. Many middle managers in large corporations, charged with the responsibility for forecasting the market for a new technological product, have probably opted for very conservative (low) forecasts because of their understandable aversion to staking their reputations and careers on a necessarily risky forecast. A key task of the technological entrepreneur—whether in a large corporation or a start-up venture—is making just such risky forecasts. In spite of the difficulties, a conscious and explicit forecast of new markets—one that draws on the best insights of marketing and other professionals within the company and the analyses of industry observers from outside the company—is an important step in selecting those product-market segments to be pursued.

Customers: An Often Overlooked Source of Innovation

Product-market choices are not made simply by gathering the company's engineers and marketeers around the conference table and urging them to exercise their creativity. New products must fulfill real or perceived customer needs or they will not sell. Changes in those needs create opportunities for new products and for the application of new technology. Customers are the obvious source of information about their own changing needs. Thus, customers should be recognized as effective innovators. By no means do suppliers have a monopoly on new product ideas, and woe to suppliers who think they do.

Customers may, of course, have a tendency to be myopic—but then so do many suppliers. Customers are particularly useful as sources of suggestions for product improvements—incremental but not fundamental changes. The more radical the new product or technology, the less likely the customers will be to conceptualize its existence and, typically, the slower they will be to use it. Most of us are reluctant to change our behavior and slow to develop a need where none previously

View customers as primary sources of product improvement innovations.

existed. The kitchen worker—at home or in a restaurant—is a source of great ideas for dishwasher redesign but is unlikely to perceive the opportunity for microwave cooking. This same person will quickly accept the redesigned dishwasher but will often be slow to accept microwave cooking.

Consider again the continuum from invention (brand new technology) to innovation (at the extreme, incremental product improvement) (Figure 2-1). The farther one operates toward the innovation end of that continuum, the more one should view customers as a prime source of new product ideas.

When the users or customers are themselves technologists, they represent a particularly productive source of new ideas, some of which may even be well along toward the invention end of the innovation-invention continuum. Erik von Hippel reports that in a study of a broad sample of commercially successful product innovations in the scientific instrument field—instruments used by technically sophisticated persons—more than 80 percent of the products sampled were "invented, prototyped, and utilized in the field by innovative users before they were offered commercially by . . . instrument manufacturing firms. This user-dominated innovation pattern remains "strikingly high for basic innovations and for major and minor innovations as well." The manufacturer's contribution to the new product in these instances was "to perform product engineering work that, although leaving the basic design and operating principles intact, improved reliability, convenience of operation, and so forth; and then to manufacture, market, and sell the improved product."[12]

Customer-technologists are a great source of innovation and even invention.

Von Hippel also comments that "user-dominated innovation appeared to hold for companies that were established manufacturers of a given product line—manufacturers who 'ought' to know about improvements needed . . .—as well as for manufacturers for whom a given innovation represented their initial entry into a product line."[13] Von Hippel also found many instances where users strongly influenced but did not dominate (according to his definitions) the technical innovation.

An extension of von Hippel's research to other industries showed somewhat less user domination of innovations in process equipment, where users are likely to be somewhat less sophisticated in the technology that is fundamental to the equipment, and virtually no innovation by users of engineering polymers and of additives for commodity plastics.

Coupling to Access Customer Innovations

A company's sales staff carries the primary responsibility for customer relationships. Competent sales managers know that an important part of their jobs, and that of their staffs, is to ascertain and communicate changing customer needs. Yet, at the same time, the larger task is to sell those products or services that the company now produces. The necessary preoccupation with selling today's products and services tends to impede the corollary task of assessing what new products or services will be appropriate in the future to meet the customers' changing needs.

Organize your high-technology business to facilitate coupling

If the customers represent a (and perhaps *the*) major source of new innovations, the systematic gathering and processing of these new ideas is important to every company. In a high-technology environment characterized by rapid change, this coupling between the customers and the company's development staff assumes critical importance.

When the marketing and the development executive is the same person—as in the case of a technology entrepreneur—coupling is easily accomplished. The entrepreneur is continually interacting wih the customers, learning and evaluating their present and changing needs from both a marketing and an engineering perspective. The entrepreneur, responsible for the firm's technical direction, knows the capabilities and limitations of the company's technical staff and facilities. He or she can evaluate new product or service ideas elicited from customers in light of the company's technical competence and future direction and make new product decisions quickly and efficiently. Indeed, the coupling within one person, the entrepreneur, of marketing and engineering perspectives gives the entrepreneurial firm a significant advantage in competing with larger companies in high-technology markets.

Effective entrepreneurs are good couplers.

In larger companies, organized along functional lines, this coupling is more difficult, but no less important. The discus-

Involve engineers in trade shows, customer visits, and users group meetings.

sion thus far has assumed that the sales organization is part of the marketing function. Where this is not the case—that is, where the top sales and the top marketing executive both report to the general manager—the coupling task is further complicated because communication lines are lengthened (although some advantages may attend this split in responsibility).

Other devices can also be used to effect coupling. Programs should be set up to permit—or cause—engineers to visit customers on a regular basis or to travel with sales personnel for a few weeks each year. Key sales personnel should be invited to technical briefings where development engineers outline their ideas regarding new technology directions and further exchange of ideas between sales and engineering personnel is encouraged. Industry trade shows represent an ideal opportunity for customer-engineer coupling. Key development engineers should typically attend such shows because a few hours or days in the exhibit booth can be a very efficient way to gather customer opinions and complaints. Trade shows also permit engineers to see firsthand the products—particularly any new products or features—the competition is offering.

Another device, users groups, originated by computer customers and now spreading to other industries, has coupling as a primary objective. Users groups were created as a kind of trade union for customers; customers share ideas but also bring pressure on their common supplier to implement certain changes in the products or services. Most enlightened suppliers recognize that these users groups can not only be helpful (and also potentially harmful) in marketing but can represent an efficient device for customer-supplier coupling.

An important argument for creating divisions and decentralizing along product lines (or along market lines) in large companies turns on the issue of coupling. Small divisions have fewer layers of management; thus, interaction among the division's general manager and technical and sales personnel is more efficient and rapid.

Product Managers

The designation of product managers (PM) represents a useful short step in the direction of divisionalization. A product manager has broad responsibilities for developing, manufacturing, and selling a product (or a product family). The PM typically has no direct line authority over any of these three functions—development, manufacturing, or sales—but is responsible for effecting coordination among them all and for setting the marketing (as opposed to selling) strategy for the product. The PM may report to the general manager or, alternatively, to one of the functional executives, typically either marketing or development.

The PM form of organization brings the coupling function to a middle management position: The PM is responsible for knowing the customers' changing requirements, for understanding the firm's technical competency, and for making new product recommendations and marketing strategy decisions in light of conditions both in the marketplace and in the development laboratory. The PM acts as a quasi-general manager and typically has a profit (or quasi-profit) responsibility.

Use a PM to improve coupling.

A common criticism of this organizational form is that the PM has a great deal of responsibility and almost no authority, inasmuch as neither the sales staff nor the development staff works for him or her. Nevertheless, the use of PMs is increasing in popularity as the need for close coupling within a dynamic environment is recognized. An advantage of this form of organization is that it creates a cadre of middle managers—the PMs—who, relatively early in their careers, are required to assume a general management perspective. To the advantage of individuals assuming PM positions, but perhaps not to their employers, the PM job seems also to represent an ideal training ground for future technical entrepreneurs who may decide to break away and start their own companies.

∴ Customers are often overlooked as a source of innovation and even invention. The importance of customer coupling depends on both the nature of the product and the position and technical sophistication of the customer. They are a more useful source of product innovations than process innovations and of product improvements than of fundamentally new products. Technically sophisticated customers are particularly useful. Therefore, extensive interaction between customer and supplier is typically desirable, and organizing to effect that interaction—that is, organizing to couple up with the user—is critical to the high-technology company.

➤ *Patents: An Unreliable Barrier*

That patents command little attention in this or subsequent chapters may be somewhat surprising, considering I am focusing on high technology. Fundamental patents have been extremely important in the success of a few high-technology companies. Xerox and Polaroid are paramount examples. But, in fact, the examples are few because relatively few patents are truly fundamental. An overreliance on patents as a barrier to entry by competitors is unwise.

First, most patents can be legally circumvented or avoided if the market opportunity is sufficiently attractive to potential competitors. Second, patents are expensive, particularly filing them in foreign countries (a necessary step for any important patent). Third, a distressingly high percentage of patents is held to be invalid (typically because of the existence of "prior art") when tested in court. Fourth, defending a patent position either offensively or defensively is expensive. Frequently, the smaller high-technology company has no practical alternative but to license its patents to larger competitors or take licenses from large companies under patents that it feels are of questionable strength.

Patents form a weak competitive barrier.

Thus, patents simply do not represent a very effective competitive weapon. "Being first" with new technology and "running fast" typically present greater barriers to competitive entry than do patents. Unpatentable "know-how" is also frequently the key. Nevertheless, most high-technology companies pursue patents on new discoveries, for the following reasons:

1. Such patents may provide some obstacles for competitors, obstacles that may lengthen the competitors' design cycle or otherwise impose some cost penalties on them.

2. The existence of patents may have some marketing advantages, implying that the company is a technological leader.

3. If a company fails to patent its discoveries, a competitor may later make the same discovery and apply for a patent, thus creating a nuisance for the first discoverer.

4. Patents represent psychological compensation to the inventor, even though the patent is owned by (and was paid for by) his or her employer.

∴ A high-technology company cannot rely on patents to keep competitors out of a market. It is more important to be the first to create and use a new technology.

Categorizing Products by Use and Degree of Standardization

Product-market segments can be usefully categorized in terms of (1) the use to which the customer puts them and (2) the degree to which they are of standard or custom design. The nature and extent of interaction between marketing and engineering at a company producing standard products are different than at a company producing products designed and built to the customers' specifications (custom products). These interactions also depend upon whether the customer is the end user of the technical product or is incorporating it into another product for sale to the end user.

The operating manager should analyze the product categories within his or her high-technology company, recognizing the difficulties of combining several categories within a single business unit. The demands on marketing and engineering—and on manufacturing—can be very different.

Analyze your company's product categories.

OEM versus End User Products

Coupling between customer and supplier is critical when the supplied item becomes an integral part of the customer's product or service. The customer in this case is the original equipment manufacturer, referred to typically as an *OEM customer* or simply an OEM. A great variety of industrial goods is sold to OEM customers, from nuts and bolts to minicomputers. In many instances, a company supplies both OEMs and end user customers with essentially the same products. When the products sold to the OEM are standard, interchangeable with those competing firms offer and based upon mature technology, OEMs do not behave dissimilarly from end user customers, except that they typically purchase in large quantities and demand and receive additional price discounts.

OEM purchasers of customized products demand close engineering links.

OEM customers of high-technology products, or products that have been customized, frequently place heavy demands on the supplier's engineering department. The supplier may be required to develop and build products to specifications tailored to the particular application and not to a set of general purpose applications. Producibility (in order to achieve cost targets implicit in the discounted price to the OEM) and reliability (because failure will reflect more on the OEM than on its supplier) take on increased importance as design criteria. Product changes must be carefully orchestrated with the OEM, who may in turn have to change its end product in order to accommodate the change. Direct communication between the engineering departments of the two companies—supplier and customer—are likely.

If the customer is an OEM rather than an end user, the role of the sales or marketing personnel is intense prior to the initial sale but substantially reduced thereafter. Once the original sale is made and the product gets "designed in" by the OEM customer, continued business is reasonably assured without much additional selling. The key is getting the product designed in. A supplier may welcome the opportunity to adapt a standard component to the OEM's particular requirements, even though such adaptation may be expensive, if the result is that the OEM becomes locked into the continued use of the component. Sales personnel will expend much effort in working with the OEM's engineers to assure that the design specifications are so drawn as to make the supplier's component acceptable. In some cases, the supplier's sales personnel may be able to get the specifications written so that *only* the supplier's component can meet them—a highly desirable result for the supplier and a precarious position for the OEM.

Get the OEM customer to design in your product.

Selling expenses are typically low for companies whose

output goes entirely to OEMs. They must be; OEM customers demand price concessions to reflect the fact that they assume the more difficult and expensive task of selling the end users. Thus, once the initial sale is made—the product or component is designed in—the sales department's role becomes primarily one of facilitating communication between the customer and the various departments within the supplier company, particularly manufacturing and engineering.

OEM customers have a particular interest in the future directions of the supplier's technical efforts and product changes. The more thoroughly the component is designed into the OEM's products, the more concerned the OEM becomes that changes in the component will trigger substantial and expensive redesign in its own products. Where relationships between OEM and supplier are long term and cooperative, the supplier provides considerable advanced notice of possible design changes, keeping the OEM abreast of future directions of development effort.

These relationships can be prickly. The OEM customer wants a great deal of disclosure and a minimum of surprises. The supplier wants to maintain confidentiality, for fear that its development plans may be leaked to competitors. The OEM, in turn, may have new product plans that must be revealed early to the supplier in order that the necessary component can be redesigned. It, too, wants to maintain confidentiality and worries that the component supplier may reveal its plans to competitors to gain a marketing advantage. An atmosphere of trust and cooperation may take years to build. Supplier and customer must recognize that they are mutually dependent. Nonetheless, each wants to maintain its independence on issues of pricing, delivery, and other terms of sale.

An inherent OEM-supplier conflict revolves around confidentiality and cooperation.

OEM customers often have a special relationship with the manufacturing function, as well as with the engineering department, of supplier firms. I explore this relationship in Chapter 3.

The Standard to Custom Continuum

In considering products, we again encounter a continuum: from standard products that are detailed in the supplier's catalog and never altered for individual customers to products that are designed and built for a single customer and never offered to others. At the custom end of this continuum, the supplier is in essence a technical consulting firm: The item delivered to the customer may be a physical product, but the company is in reality in the service business—supplying engineering design services—rather than in the product business. Custom computer software firms fall at this end of the continuum.

Moving toward the other end of the continuum, we encounter many technical companies who provide more or less standard products that are customized to a particular buyer's application. Much process equipment falls into this category, as does packaging equipment and certain control devices or systems. Computer software firms that offer standard programs adapted to individual customer requirements also fit here on the continuum.

Farther along toward the standard product end of the continuum are suppliers of products or systems that are configured to customer requirements but from standard components or major subassemblies. Computer system suppliers, manufacturers of certain heavy capital equipment (such as printing presses or machine tools), and suppliers of sophisticated analytical instruments are examples.

Resist the temptation to operate across the continuum.
Many technical businesses face the almost irresistible temptation to operate across much of this continuum, from standard products to custom products. For example, a manufacturer of standard motion transducers may be tempted to produce special high-priced transducers for aerospace applications, and the producer of specialized telecommunication equipment for the military may be intrigued by apparent opportunities in the commercial PBX market. Operating at the standard end of the continuum requires marketing to pass up sales opportunities that require customization and engineering to focus on developing new products with broad applications. Operating at the custom end of the continuum implies pricing at a level that eliminates certain market segments and requires a great deal of engineering depth and strong bidding capability. The temptation is to offer either custom adaptation of standard products at inadequate prices or custom-designed products at standard-product prices in the (gener-

ally vain) hope that the same products then can be sold to other customers to recoup the cost of the special engineering.

Seldom can the efficient design and manufacture of standard products be mixed in the same organization with the custom design and fabrication of special products; the demands on all of engineering, marketing, and manufacturing are simply too different. Note the number of military contractors who have been frustrated in their attempts to enter commercial markets. Technical companies are well advised to select a product strategy somewhere along the standard-custom continuum and then resist the temptation to move too far afield.

The High Risk of Customs. Some companies have been successful at the custom end of the continuum, but many more have not. The risks are high. The profit from many successful custom-design projects can be dissipated on a single project that proves to be substantially more difficult than was foreseen when the project was originally bid. The customer's requirements and operating environment must be understood thoroughly; sales personnel must be technically qualified; and those responsible for bidding must allow for contingencies and a great deal of customer hand holding when arriving at a final price.

Although pricing on a cost-plus basis can eliminate a large part of this financial risk, such an arrangement normally also limits the profit opportunity. Moreover, engineering resources, frequently the scarcest of all resources in a high-technology company, must be extensive if the company is to be successful in the custom business. Virtually all technical companies offer their customers some mix of product and service, but at the custom end of the continuum, this mix is heavily weighted toward service.

Custom products have a heavy "service" component.

Product Families. Many technical companies have, through carefully designing their product families, successfully achieved some of the economies in engineering and production costs associated with standard products, together with the higher prices associated with products that are tailored or customized to customers' particular requirements. By offering a set of standard options, the company may be able to provide its customers with a wide range of performance capabilities while incurring only minor design costs on each sale. The customer obtains a product whose final configuration is tailored to its particular need. To the manufacturer, the product is standard, having been configured from components or subassemblies that are available in inventory. Thus, a process-control system can include standard sensors, input-output devices, computers, memory, and software designed and built for stock but then configured into a final system in accordance with the customer's needs. A computer system can consist of standard central processing units, memory devices, printers, input devices, display units, and operating systems; yet no two complete systems the company delivers may be the same.

Product families can permit economical semicustomization.

Careful attention must be paid to the design of these standard modules. The temptation is to proliferate modules in order to meet the exact requirements of a broad range of customers. Designing, manufacturing, and inventorying fewer modules may offer some real economies. The result may be that some customers are supplied with more capability than they need or have purchased, but customers are unlikely to complain about this eventuality. It may be possible to disable certain features of the system that the manufacturer has elected to supply but the customer has not paid for.

Avoid proliferating members of the product family.

When a range of products is to be offered to the market, careful marketing and engineering analysis is required to design an optimum set of product specifications and prices. Ideally, from the supplier's viewpoint, each customer would be charged the maximum that he or she is willing to pay, given the customer's particular application. Although such a pricing strategy can be attempted with custom products, it is impractical for standard products. Moreover, it is illegal because price discrimination among like customers is forbidden in this country.

A careful analysis of the use to which the product will be put by each of the customer segments may suggest a set of product offerings at prices that will be optimum for the supplier. Consider, for example, a line of specialty pumps that finds application across a variety of petrochemical and other industries. Assuming these pumps result in lower operating costs for the customers, the supplier should seek to charge a price for each pump just low enough to make its purchase attractive under the customer's capital expenditure analysis procedures. A higher price will eliminate the sale and a lower price will be suboptimal for the supplier. Not all customers have the same return requirements for capital investment projects, nor will each application have identical cost reduction opportunities. Nevertheless, a kind of demand chart that arrays the operating cost savings (and thus rate of return opportunities) by customer segment can be constructed. A family of pumps with increasing capabilities at increasing prices may be offered to optimize returns from this set of standard products offered to a variety of customers.

Recognize pricing opportunities within the product family.

Customer-Funded Development. Most technical companies fund development activities from their own resources, expecting these expenditures to be more than recouped from the subsequent sale of the standard products. Those companies operating at the extreme custom end of the standard-custom continuum expect each customer to recompense them fully for the design of each product. Occasionally, however, the opportunity arises to obtain customer funding for a development project that can lead to one or more standard, proprietary products that can subsequently be offered to other customers. Both sales and engineering staffs should be alert to these opportunities.

Be alert for opportunities to have customers fund product development.

A manufacturer entering into one of these contracts needs to limit carefully the rights that the customer who is funding the project retains in the resulting product. The manufacturer should be certain that the new development project is consistent with its product and technical strategy and is not unduly disruptive to the engineering department's existing agenda.

Except for those technical companies whose primary market is the military, these opportunities for customer-funded development are scarce. It is an unusual situation when a customer, operating in a competitive industry, will fund the development of a new product by a supplier who is then free to offer the product to the funding company's competitors. Sometimes, however, the right of exclusive use of the new product for a year or so may be sufficient inducement for the customer to fund the development. Companies whose primary activities are at the standard end of the continuum, however, should guard against the possibility that a series of these opportunities for customer-funded development will cause it to drift unwittingly toward the custom end.

∴ Whether the customer is an OEM or an end user and whether the product is standardized or customized affect the interactions at the engineering-marketing boundary. Careful and imaginative product design may permit a supplier to gain certain of the marketing and pricing advantages of custom products without suffering the cost and time disadvantages generally associated with individually engineered products or services.

◆ *Application Engineers: At a Critical Boundary*

In this chapter, I have focused on the working relationships between engineering and marketing within a technical company. At the boundary between these two primary functions are application engineers, who are responsible for assuring that the company's products or systems are specified in an optimum way for the customers' particular applications. The application engineering activity is particularly critical when the product or system offered contains both standard and custom elements or when the system is configured from an array of standard modules. Configuring the system and setting the specifications for the custom elements are the job of the application engineer.

The best application engineers are both excellent sales-persons and strong engineers. Their first task is to understand thoroughly the customer's environment and needs. This understanding often cannot be gained solely by discussions with the customer's staff because these individuals may be so close to the immediate problems that they are unable to visualize the range of possible solutions. Thus, the application engineer must be a diagnostician, seeking out true requirements of the customer's application or the root cause of the customer's problem. He or she also needs to understand the limits and range of the customer's operating conditions, the accuracies demanded, speed requirements, probable frequency of preventive maintenance, technical sophistication of the customer's operating personnel, and all the other factors that can influence product or system final performance. Throughout this investigative process, the application engineer should be alert to opportunities to reinforce the selling effort.

Application engineers must be both salespersons and engineers.

The application engineer must have thorough knowledge of the products that he or she represents, as well as of the capabilities of the design engineering staff. The more sophisticated the technology of the products and the more the products are to be customized, the more likely that the application engineer will need formal engineering training.

Typically, the application engineering function reports to marketing because coordination between the sales staff and the application engineering staff, as each focuses on the customer, is so critical.

Application engineering activities should not be shortchanged. Many technical companies, including IBM, Foxboro, and Hewlett-Packard, utilize application engineers aggressively, recognizing that much of the value delivered to the customer is, in reality, application engineering services. When these services are outstanding, the customer will be willing to pay for them in the form of higher prices for the products or systems delivered. Too many financial controllers look at application engineering as just another overhead expense that should be minimized. Application engineering expenses, like selling expenses, need to be managed but not necessarily minimized. Well-performed application engineering can result in higher prices and greater profits in the short term, to say nothing of more satisfied customers and repeat orders over the long term.

Do not view application engineering as just another expense to minimize.

Application engineering is a good entry-level position.

Application engineering is an outstanding entry position and training ground for individuals aspiring to careers in either design engineering or selling. Future design engineers benefit from the exposure to customer's needs, foibles, and operating environments. Future sales engineers benefit from gaining a thorough knowledge of the company's product line and engineering capabilities. Young persons entering high-technology industries should recognize application engineering as a desirable first position. Companies, in turn, must promote promising individuals rapidly from application engineering into design engineering or field sales in order to make application engineering a good career-building step.

∴ Application engineers add important values to the package of products and services offered to customers. They must be both effective sales representatives and technically competent engineers.

Selling Technical Products: By and to Engineers

Selling technical products differs from selling more mundane products and services, largely because it involves selling both by engineers (or at least technically trained personnel) and to engineers. The sales staff in high-technology companies is likely to be heavily populated with engineers. If the technical product is to be sold to engineers—for example, electronic test equipment or robotic systems—the sales staff may be comprised entirely of engineers. When the product offered is highly technical but the customers are not—for example, word processing equipment, certain medical electronic devices, or laser aligning devices used in construction—the need for trained engineers within the sales force is sharply reduced.

In selling any products to OEMs and custom products to end users, the firm's engineering staff becomes directly or indirectly involved in the selling process. A customer who seeks unique or state-of-the-art technology and who is concerned about the impact of technological changes wants direct access to the supplier's engineering staff. In these instances, the engineering staff can be as important as, or more important than, the sales staff in effecting the final sale. The need for effective coordination and harmonious relationships between the two staffs is obvious. Top management should be as certain that the engineering staff is trained in good selling techniques as they are that the sales staff is technically knowledgeable.

What, Who, and Why

Two fundamental questions about the customer are no less applicable to the sale of technical products than to the sale of other products. However, the answers to these two questions may be substantially more complicated when the product or service is technically sophisticated.

The first question is what is the customer buying. The question does not concern what the customer *says* he or she is buying or what the salesperson is selling. Rather, the important question is what set of personal and corporate needs is to be fulfilled by this purchase. Prestige, reliability, personal security, product appearance, or other intangibles may be every bit as important as speed, mean time between failures, capacity, energy consumption, accuracy, price, and other properties prominently displayed on the product's data sheet.

Know what your customer is buying and who makes the decision to purchase.

The second question is who buys and why do they buy. Who are the real decision makers? Often in the purchase of technical products or services, several individuals or functions in the customer's company will have a say in the final decision, for example, design, manufacturing, facilities engineering, purchasing, quality assurance, and manufacturing management. Who are the key decision influencers and what is the relative importance of each in the final purchase decision? What are the buying motivations of each—there is likely to be quite a range of motivations—and how can these motivations best be influenced?

The sale of technical products to rational, technically sophisticated customers is frequently assumed to be free of the emotion that pervades the sale of nontechnical products to consumers. This is a risky assumption. Engineers are probably as, if not more, susceptible to emotional decision making as consumers in the supermarket. Although they may analyze the physical properties and price of the technical product with sophistication and rigor, the product comes wrapped in a set of intangible properties not so susceptible to rational analysis (such as the supplier's reputation regarding technology leadership, fair pricing, operator training, on-time delivery, and obsolescing of products). Another key parameter is the salesperson's ability to make the purchaser comfortable with his or her decision. Effective sales presentations and closing the order· are essential; even products incorporating superior technology do not sell themselves.

Emotion can play a great part in an engineer's decision to buy.

Thus, the salesperson faces a range of possible buying motivations, some emotional and some economic. Some products or systems, such as process control or energy management systems, offer the customer immediate and demonstrable economic savings. Others, such as robotics, computer-aided design and manufacturing systems, or office automation systems, may offer enhanced capabilities and the potential—often very difficult to quantify—for future cost savings. Others offer to OEMs features that enhance their product's competitive position. After-sale service, enhanced reliability, or the personal security that attends purchase from the recognized industry leader may be strong motivations for certain purchasers. Crisp industrial design plays no less a part—and arguably a greater part—in the sale of high-technology products than of more mundane products. Finally, "sizzle" or "pizzazz" also sells many technical products. One wonders how many computer systems, corporate jet aircraft, numerically controlled tools, communication systems, or automated warehouses have been bought, not for their tangible benefits, but for the prestige, excitement, or ego gratification they have brought to the key decision makers or top managers.

Sizzle and pizzazz help sell high-technology products.

Wise sales managers and product designers—that is, both the marketing and engineering departments—should bear in mind this broad range of buying motivations as they develop the product specifications, product packaging (appearance), and sales aids for the field force.

Sales Staff and Innovation

Conventional wisdom holds that innovation is the responsibility of engineering, and the sales staff's responsibility is to sell what engineering has innovated. Earlier I suggested that, in fact, customers or users represent a, if not the, primary source of product innovation. Although most companies seek various devices to couple their own and their customers' engineers in order to capture these customer-originated innovations, the sales staff represents the primary link. The best salespersons remain alert to possible new products or product improvements that may come to light in the course of conversations with customers. Sales engineers themselves, intimately involved with customer needs and requirements every day, should represent a valuable source of new ideas. Well-managed technical companies see to it that sales engineers are trained to recognize important new ideas and that new product ideas originated by the sales staff obtain a proper hearing in engineering and are not dismissed as a result of the all-too-prevalent "not invented here" attitude.

Salespersons should be always on the lookout for new product ideas from customers.

Pascale and Athos, in a book on Japanese and American styles of management, report that companies with the most loyal employees are likely to be the most innovative.[14] I have already commented on von Hippel's findings that most product innovations come from customers. Combining these two observations suggests that loyal and stable sales forces—ones whose interests are long term and extend beyond sales commissions checks—may be particularly able to assess and transmit customer-generated new product lines. All of this complicates the task of sales management in a technical company. The sales staff must be oriented both to the effective selling of today's products (without becoming preoccupied with how the products should be changed) and to the identification of new product opportunities. Sales training programs, compensation schemes, and promotion policies influence both how the sales staff responds to its dual role and how stable and loyal it is to the company's long-term objectives and goals.

Channels of Distribution

The choice of channels of distribution for technical products is generally quite constrained. Technically sophisticated products and services are not very susceptible to indirect selling. Dealers or retailers have traditionally played essentially no role. (Recently, however, some computer and office automation equipment has been offered through retail outlets, some of which are owned by manufacturers such as Texas Instruments and Xerox.)

Should you use a direct sales force, agents, distributors, or a combination?

The classic choice facing sales managers is among a direct sales force, agents (manufacturers' representatives or "reps"), distributors, or some combination of the first three.

The distributor's primary functions are to carry inventory, service the customers' purchasing and inventory control departments, and extend credit. When rapid and frequent delivery of standard parts from a local inventory is important to the customers, distributors play a vital role. Few distributors employ technically qualified salespersons; thus, the distributor's sales force can typically offer customers little technical information or assistance beyond that contained on the product data sheets.

The dilemma many smaller technical companies face is whether to deploy their own sales forces or engage a network of agents or reps. Developing and maintaining a field sales force is expensive and represents largely a fixed cost, independent of sales volume. Reps, however, are paid only when they make a sale at commission rates that vary between about 5 and 15 percent. Thus, commissions paid to reps are a variable expense. Nevertheless, after the start-up phase, total commission expenses may be greater than the cost of a direct sales force. Lowest commission rates (about 5 percent) apply to standard products sold in large quantity to OEMs. Fifteen percent commissions are paid on technical capital equipment sold to end users. Still higher commission rates may apply to custom-designed products where the rep is performing key application engineering services.

By comparison with a network of reps, a direct sales force can typically be better trained and therefore can provide the customer with more technical assistance. The direct force has a single mission: to sell the company's products. Reps have divided loyalty and must spread their time and attention over a number of manufacturers' product lines. It is unusual for the reps to know the intricacies of the manufacturer's products in the same depth as members of a direct force. Moreover, the reps must be trained, assisted, retrained, and motivated by the manufacturer's own personnel; a rep neglected by his or her principal will not be effective. Capable reps have a knowledge of the territory that would take members of a direct sales force years to acquire.

An obvious strategy is for a new technical company to use reps during the early years of its life, switching to a direct sales force as it gains financial strength and greater knowledge of the marketplace. This strategy is not so easily implemented, however. The mere hint that a manufacturer may "go direct," eliminating the use of reps, will demotivate the reps to the point of uselessness. Indeed, many reps demand contractual protection for a period of years to guard against the manufacturer going direct in the particular territory and thereby capitalizing on the missionary sales work the rep has done. Moving to a direct sales force is a form of vertical integration—in this instance, forward integration. Chapter 4 discusses at some length the pros and cons of vertical integration in high-technology companies.

Switching from reps to direct is appealing but perilous.

Promotion

Certain promotional devices that are ineffective for low-technology products can be very productive for high-technology products. And some of these devices have a very low direct cost to the manufacturer.

The trade press will typically pick up press releases on new products and give them prominent display. If the journal or periodical is devoted to a high-technology industry, new product announcements are sometimes treated as news. The resulting news story represents free advertising for the manufacturer. Despite some protests to the contrary, however, a publisher is more likely to provide prominent news coverage of new products from its major advertisers.

Concentrate on opportunities for free and low-cost promotion.

Direct mail can be particularly effective where the market segment is both narrow and well defined. The manufacturer of process control systems for the paper industry can easily develop and effectively use a list of managers and technical directors at all paper mills in selected geographic regions or countries. A computer-based litigation management system for large legal firms—a definable set of customers—may also take advantage of direct mail promotion.

Technical articles written by the manufacturer's engineering personnel or papers presented at technical conferences can be another effective promotional tool. If a company seeks to convey the image of technical leadership, its technical staff's domination of industry technical conferences or publications can lend credence to that image. Such activity can also bring professional recognition and prestige to the scientists and engineers.

In a fast-moving industry, trade and technical shows take on considerable importance. When technology is evolving rapidly, customers are likely to attend these shows for more than social reasons: They need to remain abreast of the new technology if they are not to fall behind competitively. Such shows can represent concentrated and effective selling forums. Furthermore, the company's exhibit will contribute (for better or worse) to its image.

Trade shows and effective product data sheets are essential.

Product data sheets are the indispensable element of the sales promotion package for technical products. They need to be timely, accurate, complete, and attractive. They represent a selling tool and also a ready reference for both customers and salespersons.

Managers of technically based companies should actively exploit the many opportunities for publicity that are available, in addition to paid space advertising. The engineering and finance departments, as well as the marketing department, should be involved in these efforts because effective communication with the technical and financial communities can often have highly desirable sales promotion overtones.

Radically New Products

When a technical company introduces a radically new product, not simply an adaptation of or an improvement on a product already in the market, it needs to focus on the extent to which it is asking potential customers to change their behavior. This changed behavior may involve the use of leisure time (in the case of color television), meal preparation (in the case of microwave ovens and food processors), office communication (in the case of electronic mail systems), or price marking or checkout procedures at supermarkets (in the case of computer-based universal bar code readers). The more fundamental or radical the new product, the greater the behavior change that has to be induced if the product is to be successful. The greater the behavior change, the greater will be the customers' resistance.

Radically new products require users to change their behavior.

What are the implications for design engineering? Design choices should heed possible consumer resistance. Perhaps a family of products needs to be introduced—one for the conservative, resistant-to-change group of customers and another for the daring. Perhaps a series of new products, each more radical than the last, introduced over a number of years, can accelerate changes in customer behavior. (Competitors' reactions must be carefully considered.)

Induce behavior changes with careful product design.

Not all potential customers will adopt the new product at the same time or with the same ease. Some will be anxious to try something new, particularly if the benefits are significant and apparent. Others will join in buying only as the perceived risk of purchase is reduced, and some foot-draggers will be very slow to adopt any new product.

Researchers at Iowa State University observed this phenomenon with respect to agricultural products (seed and fertilizer). They identified and named five groups of consumers and indicated the percentage of the total buying population falling into each group (see Table 2-3).

Table 2-3.   Customer segments for radically new products

Group Name	Percent of Total Customers	Cumulative Total Adopting
Innovators	2.5	2.5%
Early adopters	13.5	16.0%
Early majority	34.0	50.0%
Late majority	34.0	84.0%
Laggards	16.0	100.0

Source: "The Adoption of New Products: Process and Influence" (Ann Arbor, Mich.; Foundation for Research on Human Behavior, 1959), pp. 1–8.

Although we should take the precise percentages with a grain of salt, the concept is useful. The implications for marketing are that different sales messages are appropriate at different times and for different customer groups.

Frequently a new product will enjoy a burst of sales as the innovators rush to try something new. Sales then slow when this risk-taking, innovating market segment becomes saturated but the early adopters have not yet been induced to purchase. Novelty and promise may represent sufficient buying motivation for the innovators, but the early adopters want some evidence that the risk takers have been successful in using the product. The early majority needs to be assured that the product is gaining significant acceptance throughout the industry. The theme for the late majority and laggards among industrial customers is that they will be at a competitive disadvantage if they fail to adopt the new product (for example, large scale integrated circuits rather than simply integrated circuits or computerized material requirements planning rather than a Kardex inventory system).

Change your promotional message as your product gains acceptance.

The challenge is to identify the characteristics of the customers who are likely to fall into each of these categories and then to design a marketing program that will be responsive to them. Some industrial companies are known as innovators in their industry and are always the first to try new products. Others are historically more conservative, preferring to adopt new products or techniques only after others have worked out the "bugs." The marketing message needs to change over time as sales attention shifts from the innovators to the early adopters to the early majority and so forth.

Communicating Technical Information to Customers

Communicating technical information to customers—both before and after the sale—is a joint responsibility of marketing and engineering. Sales personnel, trade shows, data sheets, space advertising, and all elements of the marketing mix focus primarily on prospective customers. But, in the case of technical products, the need to communicate typically does not

end with the sale; postsale communication of technical information to customers can be key to delivering satisfaction. Only
customers who are using the technical product effectively will
be repeat buyers, and only satisfied customers will provide
the word-of-mouth endorsements and positive references so
essential to future sales.

*Maintain postsale
communication of
technical
information to
customers.*

    The same data sheet given to the prospect is now going to
be used by the buyer as a guide to the performance that he or
she will demand from the newly purchased product. Inaccurate or overly optimistic product specifications will complicate
the problem of delivering customer satisfaction.

Costs of Overpromising Performance. A common lament of
customers buying custom (or even adapted) products is that
the sales staff tends to overpromise performance specifications in the course of closing sales. The same criticism is heard
from the engineering and manufacturing functions within high-
technology companies.

*Guard against
overpromising
product
performance.*

    The costs of this overpromising vary widely, from almost
nothing to corporate disaster. The sources of the costs are one
or more of the following:

1. Application and design engineering time. The opportunity
   costs here may be particularly severe; design engineering
   time devoted to a single customer order will be unavailable
   for projects aiming at new products.
2. Manufacturing and quality assurance time. Excessively
   demanding specifications are likely to require additional
   "tweaking" by manufacturing personnel and extra time in
   final checkout and test.
3. Field service time. Extra tweaking will also be required
   during installation, and warranty repair and adjustment
   is likely to be extensive. As these activities stretch out in
   time and the customer becomes increasingly annoyed, the
   sales force will also need to spend time reselling the customer if the sale is not to be lost.
4. Customer ill will. A company that overpromises its customers soon loses credibility in the market.

5. Sales returns. If the product never meets the promised specifications, the customer will likely return it. Its return value to the manufacturer will depend on the extent to which it had been customized.

6. Law suit. Blatant and willful overpromising may lead to costly legal action, particularly if the technical product is incorporated in the customer's products or production process.

To guard against these costs, technical companies should require that any order involving nonstandard specifications— and all custom orders—be reviewed and approved by engineering before a final (and legally binding) order acknowledgement is sent to the customer.

Customer Manuals. Technical manuals, describing the setup, operation, and maintenance of the technical equipment, are fundamental to the postsale communication for most products. Too often technical manuals are not given the care and attention they deserve. They may not be available until months after a new product is shipped, leaving early customers to figure out for themselves (or with the aid of expensive sales or service personnel supplied by the manufacturer) how to operate the equipment.

Another common customer irritant is a technical manual that fails to reflect the particular features or engineering changes incorporated into the unit actually delivered to the

Accurate, useful technical manuals are essential adjuncts to most technical products.

customer. Engineering change orders may be carefully controlled in engineering and coordinated with manufacturing (as I discuss in Chapter 4), but their impact on the company's technical manuals must not be forgotten or ignored. When a customer's system is configured from standard modules or when engineering changes are occurring with great frequency, the technical manual should be assembled in loose-leaf form. The proper pages—and only the necessary pages—can then be brought together into a technical manual for each customer.

The development and control of technical manuals is a fussy and expensive task. Inaccurate or incomplete manuals do not affect the inherent capabilities of the technical equipment, but they definitely influence the customers' satisfaction. A major competitive advantage of the Apple II personal computer over its competitors during the early years of that industry was a superbly designed and readable (even humorous) set of manuals for its customers. Xerox has devoted substantial attention to the minimanuals that are attached to its office copying equipment. These manuals, which must communicate to a wide variety of users with different levels of patience and frequency of use, are seen as key to customer satisfaction. Xerox knows that copying machines may have many ingenious and sophisticated features, but if these features are not both apparent and accessible to the users, customers will ascribe little value to them.

Customer Training. Printed technical manuals are necessary but may not be sufficient. Many suppliers of technical equipment and systems find it necessary to provide formal training programs to customers. In some instances, multiple training programs may be required—for example, separate programs for supervisors, operators, and maintenance personnel. Suppliers of office automation equipment are engaged in a major way in training customers, so are computer manufacturers and virtually all suppliers of sophisticated capital equipment and complex computer programs. This training may be most effectively performed at the customer's site or at the manufacturer's offices, depending on the number of people to be trained, the requirement for special facilities, and the opportunity and need for after-sale selling.

Customer training represents both a necessity and an opportunity.

Customer training, like application engineering, falls right at the boundary between engineering and marketing, although typically it, too, reports to marketing. Training provides an opportunity to solidify and reinforce the selling efforts, but it must convey technically accurate information that the company is willing to stand behind.

Training prospective customers is effective selling.

In some markets, where technology has evolved more rapidly than customers' understanding or appreciation of it, industry leaders use customer training to stimulate primary market demand. For years, IBM has conducted executive seminars—in effect, training of prospective customers—to expose top managers to the potential uses of computer technology. A leading supplier of sophisticated analytical instruments determined that a significant limitation to the growth of its market was the dearth of technicians trained in operating its own and related analytical instruments. As a result, it formed a training "institute" where technicians could learn to operate a range of analytical instruments. Of course, throughout such seminars or training programs, the manufacturer has ample opportunity to extol the virtues of its equipment.

A related marketing ploy is to provide, at a reduced price or even as a gift, technical equipment to schools and other independent or government-operated training programs so that their students develop a familiarity with (and therefore probably a preference for) the manufacturer's brand of equipment.

Signaling New Technology to Customers

The timing and extent of communication of brand new technology to present and prospective customers is a delicate matter. Manufacturers of products or systems that are subject to rapid technological obsolescence must decide when and how they should signal to their customers the impending introduction of new products based upon new technology.

A trade-off needs to be struck among all the factors listed in Table 2-4 as a high-technology firm decides when and in what manner it will reveal to its customers, and thus to its competitors, new product plans.

OEM customers will press the manufacturer for advance notice; the OEM seeks time to adapt its own products or services in order to incorporate the manufacturer's new product. *Balance good customer relations with the need for secrecy.* Timely adaptation by the OEM is very much in the manufacturer's best interests as well, assuming the manufacturer is anxious to begin selling the new product to the OEM customer as soon as it is introduced. Maintaining security regarding new product plans is difficult enough when knowledge is restricted to personnel within the manufacturing firm, but it becomes doubly difficult when customers are given early briefings on these plans. In light of the high rate of mobility of

Table 2-4. Advantages of early and late announcements of new products and technologies

Early Announcement	Late Announcement
Give OEMs time to adapt to new product	Give competitors less time to react
Delay customer decisions, if company will follow competitors into the market	Dramatically surprise the market (and competitors)
Dissuade competitors from investing in development, in light of company's lead time advantage	Minimize impact on sales of present products to customers who might otherwise await new product
Convey impression to market of technological leadership	Retain flexibility to adapt to early announcements by competitors

professionals in high-technology industries, new product plans are all too quickly leaked to competing firms.

When a firm expects to be first in the market with a significant new device, it may want to surprise the market with its introduction in hopes of catching the competition napping. It must carefully avoid premature signaling to customers. In other circumstances, however, the firm may wish to make an early announcement of a radically new product—that is, announce the product well before the time that it can be delivered—even if it expects to be first to the market. Such an early announcement may dissuade potential competitors from investing development money in a similar product in light of the lead time advantage that the announcing firm evidently enjoys. Early announcement may also enhance the firm's image as the leader in technology.

Carefully evaluate probable competitive responses to early signals.

Such an early announcement may also be advisable when the firm will follow one or more of its competitors to the market with the new product. In this case, early announcement may induce some potential customers to wait and compare the announcing firm's offering rather than placing an early order for the competing product. Such early announcements of new computer models by IBM has brought forth charges of unfair competitive practices. Of course, any early announcement may have the unfortunate effect of stalling sales of one's own existing (substitute) products because customers are reluctant to purchase products that will soon be rendered obsolete by the newly announced product.

∴ The importance of emotion, packaging, and various intangibles of product and company characteristics should not be underestimated in selling high-technology products. Determining exactly what the customer seeks and who will most influence the buying decision are particularly challenging and critical. The choice of distribution channels depends on the nature and technical sophistication of the product, the concentration of customers, the day-to-day servicing the customer requires, and the size and financial wherewithal of the company. Successfully introducing a radically new product requires concentrating on changing the behavior of members of the first two of the five groups of buyers—innovators and early adopters. Communicating technical information, both before and after the sale, is a vital joint responsibility of engineering and marketing that can be accomplished through technical manuals and customer training. Signaling new technology affects the actions of both customers and competitors. These actions need to be carefully forecast and evaluated.

Highlights

- The key functional interaction in high-technology companies is at the marketing and engineering boundary.

- The most critical joint responsibility of these two departments is selecting the product-market segments the company will pursue.

- The relative roles of marketing and engineering in new product definition change along the continuum from research/invention to development/innovation.

- Technological evolution is subject to technology-push and demand-pull.

- Forecasting undiscovered technology, as yet nonexistent markets, and competitors' actions is difficult, but systematic procedures are a major help.

- High-technology companies must organize and manage both their marketing and engineering efforts to take full advantage of the key role customers can play in conceptualizing and transmitting new innovations.

- Product segments can be categorized as OEM or end user products and standard or custom products.

- Imaginative product design can gain the marketing and pricing advantages of custom products without their cost and time disadvantages.

- At the boundary between engineering and marketing are application engineers, who add important value by assuring that the company's products or systems are specified for customers' particular needs.

- Just as marketing contributes to product innovation, engineering can play a key role in selling, particularly in promotion and in communicating technical information to customers.

- Selecting appropriate distribution channels requires a compromise among cost, technical competence, and proximity to the customer.

- Marketing high-technology products requires careful attention to buying motivations and influences and the extent to which buyers must change their behavior.

- Deciding when, how, and to what extent new technology and products should be revealed necessitates a trade-off among competitive posturing, image building, and possible erosion in sales of present products.

Notes

1. H. Igor Ansoff and John M. Stewart, "Strategies for a Technology-Based Business," *Harvard Business Review* (November–December 1967), pp. 71–83.

2. See Edwin Mansfield. "How Economists See R&D," *Harvard Business Review* (November-December 1981), pp. 98–106.

3. William Page, "Long-Term Forecasts and Why You Will Probably Get It Wrong," chap. 26 in Spyros Makridakis and Steven C. Wheelwright, eds., *The Handbook of Forecasting* (New York: Wiley-Interscience, 1982), p. 449.

4. James Brian Quinn, "Technological Forecasting," *Harvard Business Review* (March–April 1967), p. 89.

5. For further discussion of these and other techniques, see Harry Jones and Brian C. Twiss, *Forecasting Technology for Planning Decisions* (New York: Macmillan, 1978), and James R. Bright, *Practical Technology Forecasting: Concepts and Exercises* (Austin, Tex.: Industrial Management Center, 1978).

6. Joseph P. Martino, ed., *An Introduction to Technological Forecasting* (New York: Gordon and Breach, 1972), p. 151.

7. Ibid., p. 25.

8. Ibid., p. 98.

9. Quinn, "Technological Forecasting," pp. 101–103.

10. Ibid., p. 103.

11. Herman Khan and Anthony J. W. Wiener, *The Year 2000* (New York: Macmillan, 1967).

12. Eric A. von Hippel, "Users as Innovators," *Technology Review* (January 1978), pp. 32–33.

13. Ibid., p. 33.

14. Richard T. Pascale and Anthony G. Athos, *The Art of Japanese Management: Applications for American Executives* (New York: Warner Books, 1981), p. 190.

Chapter Three

Coordinating Marketing and Production

This chapter covers the following topics:

- Forecasting: Once Again a Key Link
- Managing the Order-Receipt-and-Delivery Cycle: A Key Short-Term Planning Issue
- What Marketing and Its Customers Should Expect of Manufacturing
- The Key Challenge for Manufacturing Management: Capacity Planning
- Field Service: At the Marketing-Production Boundary

What production builds, marketing is expected to sell. Nothing happens until a sale is made. Such are the myopic views of too many marketing people. What marketing sells, production is expected to make. Production is the key value-adding activity of the company. Manufacturing folks are equally prone to myopia. Add to these narrow viewpoints the fact that manufacturing and marketing departments are often staffed with people with different personalities—having different value systems, responding to different motivations—and the potential for conflict between these two key functional areas of the business becomes sizable.

In contrast to the marketing-engineering interface discussed in Chapter 2, where long-term issues predominated, marketing-production coordination and cooperation—or conflict and confusion, as the case may be—focus on shorter term issues involving present and potential customers for existing products. Sales forecasts and production schedules must be rationalized, and, in many high-technology companies, communication between sales and production staffs must continue throughout the order-receipt-and-delivery cycle.

Conflict regarding these short-term issues can be substantially mitigated if marketing personnel (and thereby customers) have reasonable expectations of manufacturing. Just what does the company expect of production? What should be production's priorities in light of the company's marketing strategy?

A clear answer to these questions will assist manufacturing executives in the key area of capacity planning. Manufacturing's strategy is substantially defined by its capacity—not just size, but also timing, design (that is, process technology), and location of that capacity. These decisions have profound effects on manufacturing costs; in turn, costs influence prices, which influence volume. But volume determines the rate at which the business accumulates experience. Accumulated experience leads to reduced costs—and so the merry-go-round continues. Our exploration in this chapter of the benefits and limitations of the experience curve reemphasizes the key links between the marketing and manufacturing functions.

In this chapter, we consider short- and long-term forecasting—sales and production plans—and the key related issues of lead times and delivery; operational issues that surround the order-receipt-and-delivery cycle and require extensive interaction between marketing and production personnel (for example, quoting and revising delivery commitments, transmitting order-specific information from customers to the factory, and committing to blanket orders); the set of priorities to which manufacturing can respond and the importance of rank ordering these in light of the company's marketing strategy; cost-volume-price considerations that influence capacity planning, with particular emphasis on the use (and misuse) of learning curve analyses; capacity planning (the set of decisions that implements the firm's manufacturing strategy); and a tangential but critical issue in so many high-technology companies—field service, an activity that straddles the fence between marketing and production.

The boundary between production and marketing is replete with opportunities for conflict—and for constructive cooperation. Some of these, together with typical but self-serving comments offered by marketing and manufacturing personnel, are shown in Table 3-1.

◈ *Forecasting: Once Again a Key Link*

Early in Chapter 2, I focused on the need for both the technologists and the marketeers to engage in long-range forecasting of changing conditions in technology and in the marketplace to guide the company's choices of product-market segments. At the boundary between marketing and production, the need for forecasting is every bit as critical. Here the forecasts, although somewhat more short term, must be sub-

Table 3-1. Marketing and manufacturing: Cooperation and conflict

Problem Area	Typical Marketing Comment	Typical Manufacturing Comment
Production scheduling and short-range sales forecasting	"We need faster response; our lead times are ridiculous."	"We need realistic customer commitments and sales forecasts that don't change like wind direction."
Breadth of product line	"Our customers demand variety."	"The product line is too broad—all we get are short, uneconomical runs."
New product introduction	"New products are our life blood."	"Unnecessary design changes are prohibitively expensive."
Cost control	"Our costs are so high that we are not competitive in the marketplace."	"We can't provide fast delivery, broad variety, rapid response to change, and high quality at low cost."
Capacity planning and long-range sales forecasting	"Why don't we have enough capacity?"	"Why don't we have accurate sales forecasts?"
Field service	"Field service costs are too high."	"Products are being used in ways for which they weren't designed."

stantially more precise. Sales forecasts and production schedules must be reconciled to strike a balance between making (and keeping) promises to customers and controlling the firm's operating costs.

Forecasting in a stable environment is a luxury seldom accorded a high-technology company. The market is dynamic, as customer requirements change and new and old competitors adjust their product and service offerings. The production environment is also dynamic, as products are added and dropped, volumes change, and manufacturing tackles the challenges associated with new technologies. High-technology companies are by no means immune to swings in eco-

Short-term but precise forecasts are required at the marketing-production boundary.

nomic activity, changes in international exchange rates, shifting political winds, inflationary pressures, changes in interest rates, and all the other externalities that affect all businesses. Short-term, intermediate-term, and long-term planning each provides input for a different set of decisions. Arriving at reliable sales forecasts and realistic production plans—and then meshing the two—is an enormous challenge, and one fraught with potential conflict. No high-technology company forecasts perfectly, and some make a real mess of it.

Short-Term Forecasting

Reconcile your short-term sales forecast with your build plan.

The key link between production and marketing is the reconciled short-term sales forecast and short-term production (or build) plan. The reconciliation process begins with the sales forecast but then must become iterative, as the constraints on the production floor are played off against the demands of, and promises to, customers. The objective is to arrive at a shipping schedule that optimizes among the inevitably conflicting desires of manufacturing (long runs, steady flow, narrow product range), the sales staff (rapid delivery to customers of a broad range of products), and the finance department (low costs, minimum inventory investment, and near-term earnings performance). The problems of reconciliation are compounded by a proliferation of product models and features, a phenomenon that occurs in many high-technology companies.

Short-term for some technology companies (for example, an airframe manufacturer) is many months, and for others (for example, a computer service firm), it is weeks, days, or hours. Some companies require extended backlogs in order to gain adequate efficiencies; others can respond rapidly. Management should set the appropriate term of the forecast—that is, the time period over which the forecast can be "firm"—as a function of the following factors:

1. In-process manufacturing time. The longer the time required to obtain deliveries from vendors and complete the in-house manufacturing cycle, the longer will be production lead times and the longer must be the short-term planning horizon for both sales and production.

2. The firm's production flexibility. A manufacturing company can take various steps in designing facilities, inventorying semifinished (long-lead) parts, and scheduling personnel over the short term to increase the firm's responsiveness to desired changes in the shipping schedule.

3. Competitors' actions and customers' expectations. If the industry norm is rapid delivery (for example, thirty days) and delivery commitments are key to customers' buying decisions, then the company's shipping schedule can be considered firm for only that period.

4. Profit margins and top management's attitude. If margins are wide, management is likely to demand some schedule revamping to avoid losing an order, even at a considerable production cost penalty; the marketing viewpoint will dominate. If margins are narrow, production efficiencies will be given high priority and schedules will not be altered for the sake of a sale; the production viewpoint will dominate.

A company can gain advantages in both inventory investment and competitive responsiveness by shortening the term of the forecast. In-process time can and should be reduced and flexibility gained by specific management action. At the same time, management must be alert to possible cost penalties that accompany these actions. Later in this chapter, I consider some trade-offs among these factors of cost, flexibility, and delivery.

Shorten the term of your forecast to gain advantages in inventory investment and competitive responsiveness.

Framing the Short-Term Schedule. The nature of the product or system defines the appropriate framework for the short-term schedule. Schedules for high-value ("big ticket") or custom products must be framed in terms of individual customer orders. Even somewhat longer range order forecasts should be similarly constructed—by customer name—with a probability assigned to receiving each order. (Think of the probability of receiving an order as the product of two probabilities: that the customer will buy and that the company will be chosen over the competition. Such decomposition of the probability assessment often imparts somewhat more realism to the

Shape your short-term schedule to fit your products and your manufacturing system.

sales forecast.) Forecasting in the consulting or engineering services business follows a similar customer-by-customer process. Because of the specificity of these forecasts, virtually all sales personnel must participate actively in preparing them.

When systems are to be assembled to order, but from standard components or modules, the sales forecast needs to be framed in terms of these modules or in terms of model numbers or product features you can translate into product modules. (The analogy in the service business is that not only must you estimate the aggregate short-term demand for service, but you must break it down by geographic region and particular skill requirement.) Such forecasting is dicey, because predicting exactly what features customers will select is an order of magnitude more difficult than predicting whether the customer will order at all. Yet production requires these data if the modules are to be available in a timely manner that will permit on-time shipment of the finished system. The benefit, of course, is that certain common modules can be inventoried to shorten the in-process production time on full systems. The objective here is to develop both a master schedule of modules and an assembly schedule in terms of individual orders of customized systems.

When standard components or devices are built for and shipped from inventory, the sales forecast is simply framed in terms of these catalog items. The coupling between the detailed sales forecast and the production plan can be somewhat looser in this case.

Decoupling Forecasts and Schedules.   Decoupling of production schedules and order forecasts occurs whenever companies respond to changing customer demands by building to and shipping from stock on reasonably short notice. The role that finished goods inventory should play in smoothing production flow, while serving customers well, is in turn a function of the following factors:

Finished goods inventory can decouple schedules and forecasts.

1. Product standardization. A cataloged product purchased by a broad list of customers can be inventoried with rea-

sonably low risk. A custom (or customized) product, or one purchased by only a very few customers, must be produced in response to specific orders.

2. Product value and inventory carrying costs. A product with low unit manufacturing costs that is sold in substantial quantities can be inventoried; the money tied up in adequate safety stocks is not excessive. (Of course, inventory carrying costs are also a function of the firm's financial structure and health. An appropriate inventory investment for one firm might be intolerable for another.)

3. Costs of missed deliveries. If back orders (delayed or partial shipments) are common in the industry and tolerated by customers, low inventory and periodic out-of-stock conditions are appropriate. If missed deliveries result in lost business and reduced customer loyalty, safety stocks must be high. Again, competitors' actions and customers' expectations are key.

High finished (or semifinished) goods inventories are too often seen as the easy way out. The looser the coupling between forecasts and schedules, the greater the investment in inventory. As we shall see in Chapter 6, carrying inventory is an expensive proposition in high-technology companies. Inventory should not be viewed as a substitute for close and effective cooperation on order forecasts and production schedules.

Don't use inventory as a substitute for coordinating order forecasts and production schedules.

From Order Forecast to Build Plan.   Order forecasts should be developed from solid information (bordering on commitments) coming from the field sales force—those closest to the customers. Field input generally requires some home office adjustments, typically to temper the necessary and fortunate optimism of the field force. Sales managers, as well as general managers, must learn to calibrate individual forecasters, applying different adjustment factors in terms of both volume and timing of orders.

The production staff constructs the production schedule, or build plan, based on short-term order forecasts and existing order backlogs. Negotiation between the production and the sales staff then ensues, trading off production cost penalties against missed commitments to customers and the company's near-term financial plan.

The final build plan must be a companywide commitment.

The final build plan and delivery schedule should be blessed by top management as a companywide commitment. A key question is just how much opportunity remains for near-term schedule changes.

Freezing Schedules. The marketing department wants to "freeze" the sales forecast and shipment schedule for the shortest possible time in order to maintain maximum responsiveness to changing market conditions and customer demands. The production department seeks a forecast that is frozen for the longest possible time in order that production can be planned and scheduled for maximum cost efficiency (long runs, few setups, minimum expediting, and built-in tolerance for schedule slippage). General management's job is to strike the proper balance.

Don't permit exceptions to the frozen plan.

A real, but sometimes hidden, danger is that no explicit agreement will be reached as to the period for which plans are frozen. Ad hoc exceptions to the delivery plan should not be permitted—for example, to meet an earnings-per-share target for this month. If they are permitted, the discipline for, and commitment to, rigorous forecasting soon evaporates. Forecasting involves commitments, and commitments imply discipline.

Japanese manufacturers have been startlingly successful at reducing the period for which a forecast must be held firm (by decreasing in-process times and increasing production flexibility). The Japanese emphasis on internal discipline—meeting commitments—also reduces the need for frequent or radical production level changes. The result has been sharp reductions in inventory investment to smooth production, but without a corresponding loss in flexibility to meet customer demands. They obtain these results not through any secret procedures or miracles, but rather by very careful and committed planning and a concensus style of management that assures intimate cooperation between manufacturing and marketing.[1]

Intermediate-Term Forecasting

Short-term sales forecasting—for periods of, say, zero to three months—is primarily for the purpose of shaping and reshaping the company's current production and shipping schedules. The manufacturing, personnel, purchasing, and financial staffs all seek a reliable intermediate-term sales forecast—for periods of, say, three to eighteen months—to make the following decisions:

1. Work force planning. In the short term, the only work force flexibility available to the company is the use of overtime. As the planning horizon moves out, decisions regarding hiring or discharging personnel must be made, consistent with the company's personnel philosophies and practices. The use of overtime is not a practical answer to a capacity shortfall that is expected to last more than several months.

    Use intermediate-term forecasts to improve work force and material planning, subcontractor scheduling, and inventory investment.

2. Subcontractor scheduling. An alternative to hiring or laying off personnel is to reschedule some work that subcontractors traditionally have done or might do.

3. Material planning. Whether or not you use a formal materials requirements planning (MRP) system, a somewhat longer range sales forecast permits the company to replan its major material purchases in an attempt to minimize cancellation charges, if earlier plans were too ambitious, or expediting costs, if earlier plans were too cautious. Manufacturing should use the intermediate-term sales forecasts to negotiate with vendors for improved prices, blanket orders, and accelerated or deferred deliveries. Imaginative purchasing may reveal still other arrangements that represent cost saving opportunities to the company.

4. Inventory investment. Improving inventory turnover and thereby reducing total inventory has become a major objective for manufacturing managers. During the past decade of high interest rates, emphasis on improved asset management (discussed in more detail in Chapter 6) has been a hallmark of many high-technology companies. Excessive inventories can result from poor production planning, but they become *necessary* if careful intermediate-term sales forecasting is absent, inadequately detailed,

or inaccurate. If top management permits marketing to avoid making sound and explicit sales forecasts, production planners must construct implicit ones of their own. If pressure is placed on manufacturing to respond to short-term changes in order bookings, the inevitable result will be high inventory levels. Left to the choice of being yelled at by top managers for failure to respond to customer delivery requirements—even somewhat unrealistic ones—and being yelled at by the company controller for excessive inventory, most manufacturing managers will choose the controller's wrath.

Macroeconomic Influences. Intermediate-term forecasts must take careful account of more than just immediate order prospects. Changes in regional, national, and international economic conditions will be felt over the time period covered by the intermediate-term forecast. For companies marketing internationally, as many high-technology companies do, exchange-rate movements may create opportunities in some countries and foreclose them in others. In recent years, currency fluctuations have turned favorable pricing positions vis-à-vis local competitors into decidedly disadvantageous ones in a matter of months. High-technology products are no longer immune to diplomatic intervention; opportunities to sell to an Eastern bloc or Third World country may evaporate quickly if diplomatic relations become strained between the United States and that country. Unfortunately, these factors are exceedingly difficult to forecast.

Watch the international economic conditions that affect your markets.

Companies selling components through distributor channels, or selling capital equipment, are particularly affected by economic swings, including changes in interest rates. As the economy starts into a recession, the classic capital goods cycle impacts new orders quickly. If the company's customers are in a relatively mature industry, and particularly one that produces commodity products, the falloff may be sharp. (Interestingly, in its ordering of capital equipment, the semiconductor industry in recent years has exhibited the signs of a mature, commodity, though high-technology, industry.)

Component parts that are inventoried at various points along the distribution channels—manufacturer, distributor, and user—may be subjected to an inventory cycle. As a user's business slows by some percentage and inventory reductions are sought, orders from the user to the distributor are reduced by a much greater percentage. In turn, the distributor reduces its orders to the manufacturer by a still greater percentage. The manufacturer of such components, therefore, may experience a much sharper drop in sales than might be suggested by the severity of the economic recession. Forecasters of products that are subject to inventory cycles should be certain that their system for gathering market intelligence permits them to monitor actual customer usage and not simply orders from distributors. This information is indispensable for anticipating both slumps and rapid acceleration in incoming orders.

Product Life Cycle Effects. Over the intermediate term, the effects of movement through the product life cycle will be felt, particularly if the pace of technology change suggests a short total life. Two phenomena need to be woven into the intermediate-term forecast: (1) growth of the total market, perhaps accelerated by the entry of new competitors or by particularly aggressive pricing or slowed by the onset of saturation (of the total market or of the innovators or early adopters segments discussed in Chapter 2), and (2) changes in market share induced by product improvements or price changes by the company or its competitors. The general S-shape of the product life cycle curve for the total industry is typically predictable, but the curve's exact height (total volume) and width (time period) are considerably more difficult to predict. The firm's own curve, as contrasted with the industry's curve, is just that much more uncertain.

Life cycle effects are particularly troublesome in short-lived products.

Long-Term Forecasting

For manufacturing, the great benefit of long-term forecasts—over periods of, say, one to five, and in some instances ten, years—is their use in capacity planning: planning the magnitude and placement, but also the nature, of manufacturing capacity. Longer term forecasts also highlight problems or opportunities that need to be addressed in the company's stra-

Use long-term forecasts as the basis for your company's strategic plan.

tegic plan, including plans regarding new technologies, new product families, and new market directions, as I discuss in Chapter 9.

Long-term forecasts must reflect life cycle effects, product substitution inspired by the company and its competitors, demographics, international expansion or contraction, and planned changes in market share. Planned and orderly withdrawal from certain markets can be a wise and profitable element of a corporate strategic plan. Thus, the long-term forecast both affects the strategic plan and is affected by it—it is indeed a fundamental first step in the iterative process of creating a strategic plan. But as the term of forecasting is lengthened, the opportunity for externalities—the economy, competition, and politics—to come to play is increased.

∴ In a high-technology company, all forecasting occurs in an environment of rapid change. A good short-term forecast is essential to lessen the tensions between marketing and production and to optimize among cost control, inventory investment and meeting customer commitments. Sound intermediate-term forecasts improve material and work force planning and subcontractor scheduling. Explicit long-term forecasts serve as the basis for manufacturing capacity planning and overall strategic planning.

Managing the Order-Receipt-and-Delivery Cycle: A Key Short-Term Planning Issue

In addition to rationalizing the sales forecasts and manufacturing plans, marketing and production need to interact on a daily basis as they seek to manage—mutually—the cycle from making delivery commitments to obtaining specific information on individual customer orders through to actual, on-time delivery.

Delivery Commitments

Who quotes firm delivery commitments to customers? Even if the product is shipped from inventory, stock outs will occur periodically, and the customer will demand a commitment as to when the back order will be filled. Marketing, with a sales plan to meet and repeat orders to be won, seeks rapid delivery, regardless of the cost associated with interrupting or accelerating production. Production, with expense budgets and the balance of the shipping schedule to meet, pleads for no costly interruptions.

The framework for making this trade-off between marketing and production is simple to state and difficult to implement. The foregone profit contribution that results from an order lost due to an unacceptable delivery commitment—or the cost of customer ill will created by a missed delivery commitment (a somewhat more difficult value to estimate)—must be compared to the added production cost that will arise from short-term rescheduling (overtime, repeated setups, expediting costs, and so forth). But a schedule reshuffle may have domino effects: The acceleration of one shipment, or the inclusion in the schedule of a late order, often delays other shipments, with attendant customer ill will.

Perhaps, with a little additional sales effort, a more realistic shipment schedule could be sold to the customer. Alternatively, with a little more ingenuity, the production schedule might be modified, without cost, to accommodate the important customer. The relevant costs are thus identifiable but not easily estimated. (A sophisticated computer-based scheduling system can help in making the cost and timing trade-offs.)

A negotiation process is inevitable. In the final analysis, a delivery commitment must be made. That commitment, transmitted by marketing, must have the endorsement of the production planning group. Many sales organizations promise the customer a more optimistic delivery date than the one committed to by production, leaving until later—when the customer has no viable alternative—the problem of mollifying the customer for the tardy delivery. Policy and practice regarding delivery commitments frequently diverge. When they do, the interworkings of production and marketing are subjected to substantial additional strain.

Good-faith negotiating is needed to make the cost and timing trade-off.

Specifics on Orders

When the company's business entails custom products, or systems configured to customer order, a problem arises: transmitting from the field (that is, the customer) the customer-specific order information that will permit manufacturing (and engineering, if it is also involved) to deliver the appropriate product or system. This transmittal of information sounds simple, but, in practice, it is not.

The first problem is the understandable tendency of salespersons, once a customer's purchase order is in hand, to move on to the next prospect. The higher the sales commission rate, the greater this tendency. In some sales organizations, the task of developing and transmitting the details regarding product or system configuration—including, for example, plant layout, power requirements, and other physical constraints, as well as the particular set of options to be included—is the application engineer's responsibility. This organizational arrangement should be strongly considered when the size of the individual order and extent of application engineering to be done can justify this split in responsibility. However, such a split does not represent a complete solution. The collection of these customer-specific details frequently brings to light other questions that require the salesperson's attention.

When practical, use application engineers to tansmit customer-specific order information.

Second, a delivery promise typically must be made before the exact system configuration has been tied down; yet the "build" schedule may be strongly influenced by the customer-specific details (for example, power surges or a dusty plant environment may require additional protection for the equipment). Delivery promises to customers should be accompanied with appropriate caveats.

Third, when production and engineering scrutinize the details of the order, inconsistencies may be discovered—for example, given the software ordered, more memory capacity may be required or option L and option Q, when used in combination, may require an adaptation to option B. The customer must be contacted and negotiations continued; both the salesperson's and the production planner's schedules are interrupted.

There is no easy answer. Awareness on both sides—of production's needs for information and the sales staff's handicaps in "pushing" the customer—is essential. The more complex the system, the more cooperation and interaction between production and marketing will be required to work out the order details, including the inevitable compromises and last-minute modifications.

End-of-Period Syndrome

End-of-period syndrome is widespread in, but not limited to, technical companies. Disruptive to the smooth working relationships between production and marketing, it involves shipment nonlinearity: the rushed shipment at the very end of an accounting period—month or quarter—of a large percentage of the total shipments for the period. Too frequently, two-thirds of the month's shipments occur in the last week of the month or an even higher percentage of quarterly shipments in the last month of the quarter.

The syndrome is particularly prevalent when the company's management is preoccupied with interim earnings per share (often because the management bonus formula is tied to short-term results) and the company operates with a small backlog of orders. Once the syndrome takes hold, it is difficult to correct. At the end of the accounting period, the production pipelines are empty, and the early weeks or months of the next period are spent refilling them in anticipation of another end-of-period push.

The costs of this uneven work flow are numerous. First, missed delivery commitments are common; the end-of-period scrambling leaves little time for orderly planning or coordinating with the sales department, to say nothing of the customer. Overstaffing in anticipation of peak loads begins to occur. Overtime hours are excessive, all at premium rates, and expediting with vendors results in price penalties. Quality suffers, and shipment of high-technology products occur with inadequate final checkout and calibration. These quality problems in turn result in excessive field repair under warranty, additional customer hand holding by sales personnel, and a slowdown in collection of accounts receivable—all real costs, although none of them appear on production department financial reports. This bunching of shipments at period end can also create severe difficulties for the field service force,

End-of-period syndrome extracts a high cost.

Avoid the syndrome by specifying a required backlog and imposing discipline and routine.

which is responsible for effecting installations. Finally, as I discuss in Chapter 6, shipment nonlinearity results in excessive investments in inventory and accounts receivable.

How can the syndrome be avoided? First, top management should be realistic in specifying the required backlog. When orders are slow, the financial consequences should be faced, rather than temporarily postponed by robbing from the backlog. Second, but equally important, discipline and routine must be built into the production organization—and into the supporting groups within marketing, such as order entry and application engineering—to achieve *interim* shipping targets, by the day, week, or month, and not simply the revenue targets that are implicit in the company's monthly or quarterly financial plan.

Blanket Orders

Use blanket orders to routinize the order-receipt-and-delivery cycle.

Blanket orders are purchase orders OEM customers issue for a large quantity of a component or product to be delivered over an extended but specified period of time, typically a year or more. They can help simplify and routinize the production-marketing interface with respect to portions of the company's business, just as they routinize the interaction with OEM customers. Delivery dates on specific quantities are not detailed on blanket orders; instead, customers authorize releases from the blanket order from time to time. Typically, blanket orders state the notice required from the customers for a specific release quantity.

What are the advantages of selling under a blanket order to OEM customers? Several accrue both to marketing and to production:

1. The seller is assured of the OEM's business over the time period of the blanket order. The competitive battle is thus lessened for a while.
2. Selling efficiency is gained: Price negotiations occur once, as do negotiations over quality, delivery, specifications, and other terms of sale. Releases against the purchase order are routine and can be handled by clerical personnel in the order entry and production scheduling groups.

3. Sales forecasting, and therefore production planning, are simplified. Components or products that are the subject of blanket orders can be inventoried at substantially reduced risk of obsolescence. Items on blanket order can be built during slow periods, and the production flow thus smoothed.

The OEM customer also benefits from blanket order purchasing. (Bear in mind that a company may be both a recipient of blanket orders from customers and an issuer of such orders to vendors.)

Both the seller and the buyer can benefit from blanket orders.

1. The OEM has typically designed in the component or product. Its engineering and manufacturing departments can be assured that specifications will remain unchanged during the period of the blanket order. The price is also fixed during the period (subject, in some cases, to escalation for inflation), and, to that extent, the OEM's selling prices and margins are protected.
2. The OEM's commitment to the seller is an important element in the total vendor relations package. The OEM can demand reciprocity, perhaps in the form of preferred deliveries, access to technology, or other direct or indirect assistance.
3. Just as the seller gains efficiencies, so the OEM's purchasing department finds that blanket order purchasing is efficient. Releases against the blanket order are a routine matter.
4. The OEM can reduce investments in materials inventories because notice periods for releases against the blanket order are typically shorter than normal lead times in the absence of a blanket order.

The disadvantage to both sides of the transaction is a certain loss of flexibility: A competing firm might offer the OEM a more attractive product or price during the period, and the seller must forgo opportunities to increase price or modify product specifications for the period of the order.

In actual practice, blanket orders offer substantially more advantage to the customer than to the seller. First, in most cases, the buyer is not irrevocably committed to accepting delivery of all parts ordered. If the OEM customer fails to take the full quantity, it may be "billed back" either for a modest

However, blanket orders are more advantageous to buyers than to sellers.

penalty charge or, more typically, for any quantity discount that was not in fact earned by the quantity actually accepted. The following example illustrates the mechanics of a bill back. The seller offers the following quantity discount schedule:

Quantity	Price per Unit
200	$75
500	70
1000	67

Billbacks are seldom a serious penalty.

The OEM customer issues a blanket order for one thousand and is thus billed at the price of $67 per unit. At the end of the order, the OEM customer has released only six hundred units and has thus not earned the thousand quantity price. The bill back will be as follows:

Corrected billing at higher price: (600 × $70)   = $42,000
Previously billed to the OEM customer: (600 × $67) =   40,200
Bill back   =   1,800

Of course, all bill backs are negotiable. If competition is very keen, the seller may not include a bill back clause in the blanket order. Even if one is included in the initial blanket order, bill back amounts may be forgiven or reduced in the bargaining for subsequent orders.

Second, at some OEMs, incoming inspection becomes substantially more rigorous, and the volume of parts returned for quality reasons increases, as the termination date of the blanket order approaches and substantial quantities remain to be released. Thus, blanket orders offer some real protection to OEMs but not much more than a loose letter of intent to sellers. As the semiconductor industry has undergone wild gyrations in volume—much unused capacity in 1970, 1974, and 1982 and critically high backlogs and stretched deliveries in some intervening years—the existence of a blanket order has become relatively meaningless in that industry. In periods of tight supply, customers double and triple order. When their businesses turn soft, customers cancel blanket orders routinely, with the result that apparently healthy order backlogs

Returns and cancellations can accelerate sharply.

at some suppliers have dissolved almost overnight. Indeed, more than one company has had negative orders for a month: Returns and order cancellations have exceeded new bookings.

∴Making reliable and cost-effective delivery commitments to customers requires negotiation between marketing and production. Inadequate or late transmittal of detailed specifications on individual customer orders heightens tensions and increases cost. Application engineers can often render important assistance here. Avoiding end-of-period syndrome requires that management insist on both adequate backlogs and organizational discipline. Blanket orders, although typically more advantageous to the buyer than to the seller, can routinize the order-receipt-and-delivery cycle.

What Marketing and Its Customers Should Expect of Manufacturing

So much for the thorny problems that arise in the day-to-day interactions between production and marketing. We run the risk of missing the forest by focusing on trees unless we step back and ask the fundamental question: What can marketing—and the company's customers—expect of the production department in a high-technology company? Ideally, manufacturing would make on-time deliveries at minimum cost, with maximum quality, and at the same time stand ready to adapt to changing conditions or customer demands at a moment's notice. These utopian conditions cannot be met; some trade-offs are inevitable.

All functions of the business, but particularly top management, manufacturing, and marketing, need to set and agree to the operating objectives for manufacturing and the priorities among them. The objectives and their rankings must be

Gain companywide agreement on manufacturing's priorities.

derived from a careful analysis of the demands of the marketplace, probable changes in products and technology, and the firm's competitive posture.

Rank cost, quality, dependability, and flexibility to define your company's manufacturing strategy.

Manufacturing has four general objectives: cost, quality and reliability, dependability, and flexibility. How they are ranked defines the company's manufacturing strategy. The ranked list determines what marketing and its customers can expect of manufacturing. It is incumbent on top managers and manufacturing managers to be certain that the ranking of these objectives is consistent with the company's marketing and competitive strategies. Therefore, this ranking of objectives will, to a great extent, dictate the relationship between marketing and production. The company's customers and its marketing staff will—and should—judge manufacturing's performance in light of the priority assigned to cost, quality, dependability, and flexibility.

Cost

Define cost efficiency broadly.

Manufacturing must have efficiency as an objective. Efficiency does not translate to lowest possible manufacturing cost, however. Unit production cost targets must be set in light of engineering's objectives and support (to be discussed in the next chapter). Products can be engineered for lower production costs, but the additional engineering effort may not be economically justified. Similar trade-offs exist between manufacturing and other functions, particularly field service, which is responsible for installation and warranty repair. Thus, a broader concept of cost efficiency—or cost minimization—is the objective.

Quality and Reliability

Set quality specifications in light of market requirements.

A strong emphasis on quality will result in different production decisions than will an emphasis on production efficiency. Although perfect quality may be unattainable, quality must receive greater emphasis in some product-market environments (for example, electronic medical instruments) than in others (for example, electronic games.)

Quality and reliability over the product's life must not be viewed as simply good or poor. Quality specifications are related to market needs and to the producer's technology and capabilities. Establishing appropriate quality specifications or standards is a multifunctional task; engineering, manufacturing, and marketing, as well as quality assurance, field service, and general management, must agree. Once those standards are established, production is held accountable for meeting them.

Thus, assigning a lower priority to quality does not imply that the sales force and the customers will or should tolerate out-of-specification products. Rather, it speaks to the tightness of the specifications themselves: the customers' (and marketing's) expectations regarding such measures of quality and reliability as accuracy, service life, mean time between failures, percent defective, and appearance.

Traditionally, quality and cost have been viewed as conflicting objectives. The prevailing wisdom was that higher quality results in higher costs and vice versa. In recent years, Japanese and other manufacturers have demonstrated that high quality can contribute to low, not high, cost. High quality results in lower rework, scrap, and warranty repair. When *Quality and cost* high quality results in greater yields, the average unit cost on *need not be* acceptable output decreases. I explore both quality and the *conflicting* quality assurance function further in Chapter 5. *objectives.*

Dependability

Dependability in this context means manufacturing's ability to meet its delivery commitments. The emphasis is not on minimizing the time period from order receipt to delivery, but on making good on delivery promises. In some marketplace environments, slipped deliveries are cataclysmic; in others, customers tolerate them. OEM customers tend to place *Dependability* more emphasis on dependability than do end users, but buy- *means meeting* ers of process equipment that must be installed on a tight time *commitments.* schedule also emphasize dependability.

Flexibility

Manufacturing's ability to respond in the short term to changed conditions or demands is termed *flexibility*. High-technology companies often demand much flexibility from their manu-

*Companies
experiencing rapid
change must
emphasize
flexibility.*

facturing groups, for example, when changes in customer requirements or delivery dates are imposed, adaptations in design (or new models) emerge from engineering, or engineering changes are rushed through to fix problems. The more fluid the product design and the more the final product or system must be customized to the specific application, the more flexibility is emphasized.

Trade-Offs

Where are the trade-offs? What must be given up when top priority is placed on one of these four dimensions? I have already mentioned that cost is widely perceived as a trade-off with quality. (I qualify this perception in Chapter 5.) Quality also trades off with flexibility. The more manufacturing must respond to last-minute design or configuration changes, the more difficulty production will encounter in achieving high quality standards. "Crash" fixes of problems tend to lead to other quality problems. A single-minded emphasis on dependable delivery within a company may remove from the quality assurance group the power to withhold shipments when they suspect substandard quality.

*Determine the
most appropriate
trade-offs for your
company.*

The trade-offs between cost and both flexibility and dependability are somewhat more obvious. A highly flexible manufacturing shop will typically be reasonably high cost. Little will be invested in tooling today for fear that a design change tomorrow will render the tooling useless. Scrambling to adapt to last-minute design and configuration changes will consume substantial manufacturing engineering and production supervision time, as well as the time of the production and inventory control staffs, all of which result in higher manufacturing costs. Substantial overhead expenses become built into companies that demand great flexibility from their manufacturing departments. An emphasis on dependability, at the sacrifice of cost, will permit both the heavy use of overtime when late deliveries are threatened and some built-in excess capacity in both facilities and personnel so that surges in demand can be accommodated while continuing to meet standard delivery terms.

A highly dependable manufacturing shop may have difficulty being very flexible. Responding to changes typically necessitates some rescheduling, and rescheduling is anathema to a production staff focused on punctual deliveries.

Examples of Different Priorities

A company producing a new, advanced, complex medical system that enjoys a high margin because of a strong patent position should rank manufacturing's objectives as follows: flexibility, quality, dependability, cost. Flexibility is prized because the system's newness and technological sophistication suggest that frequent design changes will be forthcoming. Naivete on the part of the sales staff as to the customers' true requirements may also result in some last-minute specification or configuration changes on individual orders. The medical application of the new system will demand that high priority be placed on quality. Dependability and cost will be assigned lower priority: dependability because on-time delivery will probably not be critical to the medical lab and cost because the high margin in the product can tolerate some inefficiencies.

Different product-market segments within the company may demand different priorities.

Suppose this same medical electronics company produces a mature line of blood-analyzing equipment, sold through medical supply distributors into a highly competitive market. The priorities for manufacturing should now be different: cost, dependability, quality, flexibility. Ranking quality below the first or second spot always seems somewhat sacrilegious. However, if any quality shortcoming in the device will result in a malfunction but not an incorrect medical diagnosis, it may not be particularly serious, especially if the medical laboratory has redundant capacity or the distributor's service department can provide rapid repair. A highly competitive market for a mature product implies that orders are won or lost on price and delivery—thus, the emphasis here on cost and dependability. A missed delivery commitment to a distributor may have repercussions well beyond the immediate loss of the sale. If deliveries are repeatedly missed, the distributor may decide to handle a competitor's line. Flexibility is not of much concern here: The product is both standard and inventoried. Because it is mature, emergency engineering changes are unlikely.

These two product lines, both involving medical electronics and the same manufacturing company, illustrate that manufacturing priorities may need to be different for different parts of the same company. In the first instance, flexibility was given highest priority and cost, lowest priority. Those rankings were reversed in the second example. This medical electronics company needs two separate manufacturing operations, each focused on the priorities as indicated. An attempt to produce both product lines in the same operation is likely to result in substantial cost penalties and missed deliveries for the blood analyzer line and confusion, delay, and frustration in attempting to produce the new, advanced technology system.

Priorities typically change over the product's life cycle.

For a particular product, these manufacturing priorities may need to change as it moves through its life cycle toward maturity. I stress later in this chapter and in Chapter 4 that, in technology-based industries, manufacturing priorities are different at each stage along the product life cycle continuum; so different production processes are appropriate. In the previous examples, the blood analyzer may have started out as a specialty, low-volume, high-technology product. At that point, it needed to be produced in the same type of production organization that the new system now demands: a job shop. Similarly, as the market for the new system matures and additional competition enters, the system's design will become more stable, volumes will increase, and price and cost considerations will assume greater importance. At that point, the production operation required will resemble the one now devoted to the blood analyzer: assembly line.

Second Sourcing

Related to the issue of dependability is the issue of second sourcing. When should an OEM customer demand a second source—that is, at least two sources of supply for a critical element in the product?

When dependability is accorded very high priority, second sourcing is in order. However, even then, second sourcing is practical only if the following conditions are met:

When dependability is a high priority, demand second sourcing from suppliers.

1. Components from two or more vendors are truly interchangeable. They have, for example, the same operating specifications and physical dimensions.

2. The OEM is purchasing sufficient volume of the component so as to be able to exercise some clout with more than one vendor.

3. The component is critical to the OEM's business, and failure to obtain it would disrupt the OEM's production and shipping schedules.

4. Delivery from the primary supplier is subject to substantial risk, typically because the supplying industry is subject to periods when demand substantially exceeds effective capacity or the probability of supply disruption is high because of labor unrest, political problems (for example, when all or a substantial amount of the fabrication is performed offshore), preemption by more-favored (for example, captive) customers, or quality (yield) problems.

An obvious corollary question is when should a supplier license (or cross-license) products to create a second source for its customers. Typically, suppliers accept second sourcing only reluctantly. Licensing others to produce one's proprietary product has the effect of reducing the product's "proprietariness" and bestowing it with more of the characteristics of a commodity product. Indeed, as noted in Chapter 2, salespersons work with OEMs to have their products designed in in ways that preclude the use of competing products.

When the customer is adamant about second sourcing and demands that a second source be available as a condition to using the supplier's product, the supplier may have no alternative. The question then becomes what company or companies to license. In the semiconductor industry, the practice of cross-licensing has become standard, and the answer to the "who" question turns on what products are available from others as part of the cross-licensing bargain. Of course, royalties from licenses can represent an important additional source of revenue. Thus, licensing between competitors can be mutually beneficial, avoiding R&D expenditures for one and generating additional revenue for the other. Furthermore, many high-technology companies may elect to license vendors to produce devices or components that have been developed in their laboratories, particularly if producing the components or devices requires expertise in which the vendors have a comparative advantage. In these cases, the licenses are more likely to represent "first sourcing" than "second sourcing."

∴ Manufacturing's strategy is defined by the order in which it ranks cost, quality, flexibility, and dependability. Trade-offs among these priorities must be evaluated in light of the maturity of the products and the technology, customer requirements, competitive positioning, and overall corporate strategy. As products, technologies, and competitive conditions change, priorities must change. As priorities change, processes must change.

The Key Challenge for Manufacturing Management: Capacity Planning

Capacity planning involves much more than just determining size.

The key challenge for manufacturing management is capacity planning: planning not just the size of the manufacturing facilities, but also the time at which the capacity will be brought "on stream," its location, and its configuration. The decisions are long term; typically, capacity can be altered only at considerable expense. Manufacturing strategy is *defined* by ranking objectives and *implemented* by means of capacity plans.

I have alluded several times in Chapters 2 and 3 to the importance of capacity planning. Long-term sales forecasting is closely linked to capacity decisions, and my discussion of manufacturing priorities pointed out the need to design (and redesign over time) production facilities to reflect the ranking of cost, dependability, quality, and flexibility. With respect to the order-receipt-and-delivery cycle, tensions are alleviated if excess capacity is available and heightened if capacity is, for example, poorly located or inflexible.

Cost-Volume-Profit Relationships

Capacity planning is really a joint responsibility of marketing and production: Manufacturing must operate the capacity and marketing must sell the output. Costs and prices must be compatible, and both are sharply influenced by volume—volume produced and volume sold.

Both marketing and production should participate in capacity planning.

A pricing truism holds that prices should be a function of what the market will bear, not of cost. Indeed, marking up manufacturing costs by an arbitrary percentage set by tradition, policy, or whim will almost surely lead to suboptimal prices. A product's value to the customer—the price he or she is willing to pay—depends upon its value in use, including both tangible and intangible returns. In a competitive market, the customer is unconcerned about and indifferent to the supplier's manufacturing costs.

Nevertheless, cost considerations obviously affect pricing decisions. Companies cannot long continue to offer products at prices yielding margins that do not cover operating costs. (This is not to say that a company might not be wise to offer products—or at least accept occasional orders—at prices below "full cost" as calculated by the accountants.)

The key linkages are that prices influence sales volumes; sales volumes drive production volumes; and manufacturing costs are very much a function of production volumes. Thus, traditional pricing practices have the causal link just backwards. Instead of costs determining prices, prices drive costs through the mechanism of volume.

Costs, in turn, are much influenced by both the size and the nature (or technology) of the capacity. Before turning to the specific questions facing the capacity planner, we must explore the relationship between costs and volume.

The learning curve links costs and volume.

Learning or Experience Curves. The learning curve or, more generally, the experience curve (the name coined by the Boston Consulting Group, a consulting firm that has at times advocated its aggressive use in pricing) links costs and volume. The learning curve proposition holds that as cumulative volumes increase, learning occurs and the costs per unit decline. One has little trouble accepting this proposition intuitively; evidence abounds that the phenomenon is very much at work in most industries.

Figure 3-1 shows a learning (or experience) curve for a producer of handheld calculators. Note several features of this representation of the behavior of costs with increased volume. First, the X-axis is cumulative volume, not volume in any particular time period. The inherent assumption is that learning is cumulative over the entire life of the product being produced. If production is interrupted for extended periods, some loss in learning will occur, but costs will not revert to the level of the first unit when production resumes. Second, the scales on the X- and Y-axes are logarithmic, that is, so-called log-log graph paper is used. The logarithmic scale reflects the assumption that any doubling results in the same percentage decrease in unit cost; that is, a doubling in cumulative volume from one hundred to two hundred units will cause the same percentage reduction in unit cost as moving from one thousand to two thousand cumulative units, even though the latter involves ten times as much incremental production. Finally, costs are inflation adjusted; that is, they are stated in constant dollars.

The first company in the market and the one with the largest market share can achieve cost advantages over competitors.

The implications of the learning curve are profound. The first competitor into a market has a significant advantage because it will accumulate experience (that is, achieve cumulative production volumes) that will provide it with a cost advantage over later entrants. This cost advantage is both tangible and quantifiable. The firm with the greatest cumulative experience to date—not necessarily the first entrant, but very likely the firm with the largest current market share—should have a significant unit cost advantage that will be manifest either in lower selling prices or higher margins.

To repeat, both the first firm to enter the market and the firm with the largest market share can have a significant cost

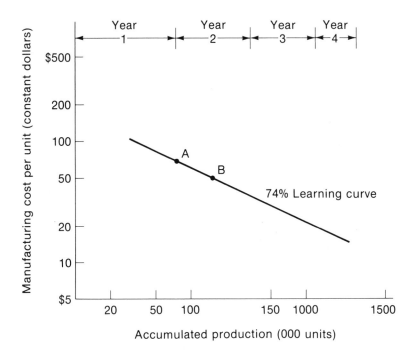

Figure 3-1. Learning curve: handheld calculators

advantage over competitors. Good evidence indicates that companies with the largest share of a particular market are on average significantly more profitable than others participating in the same market.[2] (This statement begs, for the moment, the question of defining just what is the relevant market over which to measure share.) This realization has led to some strategic planning models (discussed in Chapter 9) that seek to take advantage of this phenomenon by, for example, building share where possible and withdrawing, in an orderly manner, from markets where a significant share cannot be achieved.

A few caveats regarding the learning curve are in order before exploring further its role in capacity planning. First, learning is most obvious and measurable with respect to direct labor-intensive operations. One of the early applications was in estimating the cost over time of assembling aircraft, a rea-

Learning can and should occur throughout the organization.

sonably labor-intensive activity that is spread out over the many years of the life of an aircraft model.

Nevertheless, although the phenomenon applies only to value added by the company, it is by no means limited to direct labor. As cumulative experience builds, engineering change orders can be implemented that permit the use of less expensive components, facilitate assembly, increase commonality of parts across various subassemblies or models, or permit automated manufacture. Tooling may be improved and new capital equipment may be proposed, purchased, and installed to reduce manufacturing costs. Experience pays off in marketing and other functions as well. As cumulative volume builds, the selling task is streamlined, less missionary effort is required, and promotional efforts and expenditures are spread over a larger volume.

Suggestions for improvements can and should come from a wide variety of sources, from assemblers on the line to the manufacturing engineering staff to the order entry clerk to the general manager. Learning or experience, then, must be viewed in its broadest possible context. Reaping the benefits of experience is the responsibility of the entire organization.

Second, not all learning results in lower production costs. Restrictions by government regulations or strong labor unions can impede the process of capitalizing on organizational learning.

The slope of the curve is critically important.

Third, the slope of the learning curve is critically important and varies from industry to industry. Moreover, the slope of the learning curve indicates the change in *average* unit cost with a doubling of cumulative volume. To illustrate, Figure 3-2 details a sample calculation from the 74 percent learning curve in Figure 3-1. The average unit cost at cumulative volume of seventy-six thousand is $68 (point A on Figure 3-1—the end of year 1). At twice that cumulative volume, one hundred fifty-two thousand (point B), a volume achieved here in just a few months, the cumulative unit cost will be 74 percent of the first amount, or just over $50. Note that the average cost per unit for the units produced between seventy-six thousand and one hundred fifty-two thousand will be less than the cumulative unit cost at the higher volume, as illustrated in Figure 3-2.

Cumulative cost at cumulative volume of 76,000 = 76,000 × $68 = $5,168,000

Cumulative cost at cumulative volume of 152,000 = 152,000 × $50 = $7,600,000

Aggregate cost of the incremental 76,000 units = $7,600,000 − 5,168,000 = $2,432,000

Average unit cost for the cumulative volume between 76,000 and 152,000 = $2,432,000 ÷ 76,000 units = $32 per unit

Figure 3-2. Calculations from learning curves

Learning Curve Steepness. Technology-based businesses have the potential for steep learning curves. Although a low-technology or consumer service business might have a learning curve slope of 90 percent, curves of 70 percent or so are not unusual in high technology. Changes in technology, and particularly in process technology employed to produce the technological product, represent a major source of learning. The greater the potential for a steep learning curve, the greater the importance of insightful capacity planning.

The potential for steep learning curves in high-technology firms demands careful attention to capacity planning.

Table 3-2 shows the dramatically different impact that experience curves of 90 and 70 percent have on the average unit cost of a product as cumulative volumes move from ten thousand to twenty thousand units, assuming that in both cases the cumulative unit cost at ten thousand units is $5. The average unit cost over the next ten thousand units in the case of the 70 percent slope is one-half the average unit cost if the company is operating on a 90 percent slope. (A word of caution: Today's unit cost [at ten thousand cumulative volume] is not $5 in each case because costs have been following different learning curves. Thus, one should not conclude that, when twenty thousand cumulative volume is reached, unit costs will be 80 percent of today's cost in the case of the 90 percent slope, but only 40 percent in the case of the 70 percent slope.) This difference begins to suggest the potential competitive advantage of operating on a steep learning curve slope and accumulating experience rapidly. It also points up how

Table 3-2. Impact of learning curve slope

	90% Slope	70% Slope
Unit cost at 10,000 cumulative volume	$5.00	$5.00
Unit cost at 20,000 cumulative volume[a]	$4.50	$3.50
Aggregate cost for 10,000 units	$50,000	$50,000
Aggregate cost for 20,000 units	90,000	70,000
Difference in aggregate cost	$40,000	$20,000
Average unit cost (for 10,000 units)[b]	$4.00	$2.00

a. This value is the product of the first unit cost, $5, times the slope, because the cumulative volume has doubled.

b. Difference in aggregate cost divided by 10,000 units.

quickly a company can lose competitive leverage if others in the industry are either operating on an experience curve with a steeper slope or accumulating experience more rapidly.

Shared experience affects your learning curve and that of your competitors.

Shared Experience. Few companies produce a single product for which the learning curve operates in isolation. Experience, or learning, is often shared over a number of products. If, in the example illustrated in Figure 3-1, the company is producing home computers as well as handheld calculators and each of these products incorporates similar integrated circuits, shared experience across both the calculators and small computers might be significant, particularly if a major part of the learning is expected to occur in process improvements in the company's integrated-circuit facility.

The phenomenon of shared experience makes it more difficult to analyze the potential for cost reductions not only in one's own business, but also in competing businesses. Some breadth in product line may be helpful in gaining aggregate experience, but diluting a company's attention over a very broad product line may limit the company's ability to achieve meaningful market shares in enough of the products. Experience may also be unintentionally and unavoidably shared among competitors as key employees change jobs within the industry. Industrial espionage can also, unfortunately, cause experience to become shared.

The Learning Curve and Pricing.   The learning curve phenomenon can be used aggressively. Manufacturers can antic-   *Use the learning* ipate the cost improvements they will realize with increased   *curve to price* volume and price to attain that volume faster. The first man-   *aggressively in* ufacturer to the market, anticipating that others may follow,   *markets that are* can price so aggressively as to discourage the others from   *elastic.* entering. A sensible competitor who understands the dynamics of the experience curve will surely be discouraged by the prices being offered by, and the lead time advantage accruing to, the first manufacturer. If the discouragement is ample, the first manufacturer may eliminate some competitors before their products emerge from development.

Many high-technology markets, particularly those catering to the consumer, are highly elastic: Price reductions can bring forth substantial increased buying interest. Such markets are particularly attractive for aggressive pricing—often called *anticipatory pricing* (that is, pricing in anticipation of costs not yet realized) or *stay-out pricing* (that is, pricing to motivate potential competitors to stay out of the market).

Such pricing tactics require both courage and careful orchestration of marketing and production plans. A courageous management is required to

- Estimate the slope of the learning curve
- Make substantial investments in facilities, inventory, and   *Don't get behind* personnel on the basis of that estimate   *in the economics*
- Sell initially at prices that are below cost   *of the learning*
- Have confidence that the market exists or can be created   *curve.* that will sustain the forecast volumes
- Believe that the market share leadership position can be captured early in the life of the product and retained

Once you "get behind" in the economics of the learning curve, it is extremely difficult—and exceedingly expensive—to catch up later.

The U.S. high-technology company most identified with experience curve, or anticipatory, pricing is Texas Instruments. Some competitors still shake their heads at the aggressiveness of TI's pricing. Several Japanese manufacturers have also been advocates of anticipatory pricing, as they have gained major share positions in the world markets for cameras, audio equipment, and other high-technology products.

Key Questions in Capacity Planning

Against this background of the relationships among production costs, sales prices, and volumes, we can address the key questions that must be answered by a sound plan for productive capacity. Although they are intertwined, I consider separately the issues of size, timing, design, and location. (I leave the financing of capacity to Chapter 7.)

How much capacity should you build?

Size. Historically, American businesses have sought to centralize manufacturing activities in order to gain the economies of scale that I have been discussing. Technology-based industries have increasingly questioned the conventional wisdom that costs continue to decline as scale increases. Companies such as Hewlett-Packard are electing to divide activities, establishing a new plant (in H-P's case, establishing a new division), when employment at a particular facility reaches a certain maximum. Different companies set that maximum at, say, two hundred, four hundred, or a thousand employees, but, in any case, at levels much below the five thousand- and ten thousand-employee industrial complexes built during the heyday of economies of scale thinking. There are several reasons:

1. Except in the manufacture of very large systems (for example, commercial aircraft or submarines) or products that require extended assembly lines (for example, automobiles), the major portion of any scale economies is reached at a reasonably small size.

2. Smaller plants are more human. An employee in a large plant often has trouble identifying with the enterprise. The person may feel about as important as the badge number

by which he or she is known and feel more affinity for the labor union than the company. In a small plant, each employee is likely to know a sizable percentage of the total work force, including the general manager. Employee loyalty and identification with the operation's goals and objectives are more achievable in a smaller scale operation.

3. Related to the second reason, communication throughout the organization is simplified and expedited. A theme running through this book is that technology-based businesses demand extensive interfunctional communication—marketing to engineering, engineering to production, and so forth. Slight scale economies in direct labor and machine utilization that may be achievable in a larger operation are quickly overshadowed by improved and more timely coordination in solving problems and transmitting technical information. That is, learning in engineering, manufacturing support, and other functions of the high-technology company—and capitalizing on that learning—may be facilitated within a smaller operation.

Beware of apparent economies of scale.

4. The company avoids becoming an overly dominant employer in the town or region. This advantage has both political and employee recruitment implications. A dominant employer tends to draw more political fire—almost inevitably negative in tone—than an employer who is simply one among several. Requests for building permits are more carefully scrutinized; growth is blamed for traffic problems and escalating real estate values; and generous support of charities and community activities is all but demanded. In addition, substantial responsibility for the economic health of the city or region falls to a dominant employer.

Multiple small plants offer important advantages.

5. Each of several small plants can be optimized in design for the particular products or systems to be produced in it. Mixing process technologies in a single large facility is typically less successful than isolating them in separate facilities. The medical electronic instrument company discussed earlier would benefit fom having two facilities: one for products incorporating state-of-the-art technology and another for more mature products.

Management should resist the temptation simply to expand existing facilities. Substantial advantages can often be gained by using multiple facilities, each of more manageable size.

Timing. Only the following unrealistic assumptions would lead us to expect that demand and capacity (supply) could be exactly matched:

When should you
build capacity?

1. Lead times associated with bringing new capacity on stream are minimal.
2. Capacity can be added in very small increments.
3. Capacities can be deleted without substantial cost penalties.

Practically speaking, a technology-based company must decide as a matter of policy whether it will build in anticipation of demand—that is, build ahead of the market—or set its building plans to lag the market. The policy may be altered over time and may vary for different portions of the company's business, depending upon competitive conditions, lead times, and the cost of new capacity for particular products or product families.

A capacity plan that by policy calls for lagging the market results in less risky capital investments. The likelihood that inappropriate or excess capacity will be built is substantially reduced in comparison to a policy that seeks to build capacity ahead of the market. However, additional operating costs, arising from extensive use of overtime and some loss of efficiency because of overcrowding, also attend this policy because production must frequently scramble to meet the market demand, which has outstripped existing plant capacity.

Lag the market
when capacity is
expensive, risks are
high, or lead times
are short.

The most significant risks associated with the lag policy are the potential loss of market share, possible interruption of learning curve improvements in cost efficiency, and costs of carrying inventory built in slack periods to meet future peak demands. Note that an anticipatory capacity building policy goes hand in hand with the anticipatory pricing policy discussed earlier. If the company's strategy is to build market share aggressively, the plant capacity must be available to meet the market demand as it develops. Moreover, new and more cost-effective capacity is typically required to sustain the learning curve cost reductions that are a corollary to that share-building strategy.

The choice of whether to anticipate or lag the market in building new capacity is determined by the relationship between the cost to build the additional capacity and the opportunity cost associated with lost orders. But the opporrity cost associated with a disappointed customer may be substantially greater than simply the profit contribution on the immediate order.

Lead the market when market share is critical, lost orders are expensive, and the learning curve is steep.

In certain technology-based process industries, such as petrochemicals or bioengineering, the investment in plant and equipment is often substantial; building in anticipation of an uncertain market may be too risky. In high-technology companies that are less capital-intensive and more labor-intensive—for example, electronic instruments assemblers—the cost of facilities is low in relation to the margin earned on products sold; building in anticipation of the market is advisable. Excessive risk aversion, resulting in delays in building capacity, should be guarded against. In a high-growth market, the opportunity costs of unsatisfied demand are simply too great.

For some companies, lead times associated with bringing on new capacity may be so short, particularly if empty space is available in the area on short notice, that capacity can be added in small increments to match the growing demand. The danger here is that a series of incremental capacity additions will simply perpetuate existing operating inefficiencies. When capacity is to be added, management should seize the opportunity to make major changes in facilities design—that is, in manufacturing process technology.

Design. To many people, the word *technology* is associated solely with products. This myopia is particularly prevalent in the electronic instruments and systems business. Those in the continuous process industries recognize that technology influences both products and processes. This broadened view of technology deserves greater acceptance.

What process technology should you incorporate when building capacity?

Again, the opportunities and tyrannies of the learning curve demand attention. Improved design of manufacturing facilities is typically essential to the maintenance of a steep slope to the company's experience curve. Equally important, failure to take advantage of new process technology while the competitors are doing so can result in a company's suffering a substantial cost disadvantage.

Just as products undergo life cycles, processes may also evolve over time. Indeed, the evolution of products and processes can and should go hand in hand. Figure 3-3 provides one useful representation of this linkage. As high-technology products move from the low-volume, custom end of the product life cycle continuum toward higher volumes and greater standardization, process technology needs to change accordingly. Producing high-volume, standard products in a batch process (job shop or disconnected line flow) will place the company in a decided cost—and ultimately price—disadvantage. Excessively high variable product costs will result. Attempting to produce a wide variety of products, each in small quantity, in a highly automated facility will result in excessive fixed factory overheads. These mismatches of product and process occur when too little attention is paid to the design of the company's capacity.

Products and processes should evolve together.

Short product life cycles and rapid technological change are facts of life in high-technology companies. I stressed in Chapter 2 that sales and engineering personnel must not only accept but seek to accelerate these changes. Similarly, manufacturing staffs must be willing and eager to accommodate process changes. Yet these changes can be difficult to effect. A team of manufacturing managers accustomed to operating a job shop—with highly skilled employees, a minimum of tooling, and limited product or process documentation—may find it difficult (or at least initially uncomfortable) to design, implement, and then manage a semiautomated assembly line or continuous process. The mix of employees, the investment in fixed assets, the role of quality assurance, and the scheduling and control procedures are likely to be vastly different. Management must recognize that certain manufacturing personnel may have to be retrained or replaced and operating procedures will almost surely need substantial overhaul when new process technologies are introduced.

Small, special-purpose facilities may offer great advantages.

Because all products are seldom in the same stage of their life cycles at the same time, a single process design may be inappropriate for the entire business. Wickham Skinner points out that "a factory that focuses on a narrow product mix for a particular market niche will out-perform the conventional plant that attempts a broader mission."[3] Calling such plants "focused factories," he admonishes managers to recognize the economic potential of smaller, special-purpose facilities and to guard against preoccupation with the conventional ration-

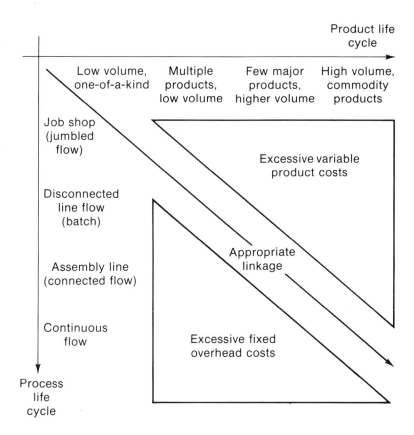

Figure 3-3. Linking product and process life cycles
Adapted from Robert H. Hayes and Steven C. Wheelwright, "Link
Manufacturing Process and Product Life Cycles," *Harvard Business
Review* (January–February 1979), p. 133.

ales of economies of scale, full utilization of existing plant
resources, and limited plant investment.

Technology advantages may overshadow experience
advantages. In my earlier discussion of the competitive cost
advantages that attend a great deal of accumulated experience, I implicitly assumed equivalent fundamental process
technologies among all competitors. A competitor can, in fact,
gain significant cost advantage through improved process
technology as well as through accumulated experience.
Attempts to segregate the two effects prove frustrating because
incremental improvements in process technology are part of
the experience phenomenon.

*Beware of your
competitors'
process technology
advantages.*

Fundamentally or radically different process technologies can shift the entire experience curve, not simply steepen its slope. For example, the cost advantage that the Japanese enjoy in automobiles, some electronic components, and certain consumer electronic devices derives more from process technology than experience. This condition is borne out by the fact that the Japanese manufacturers achieved a cost advantage over their U.S. competitors at a point when their accumulated experience was considerably less.

Managers must guard against smugness regarding process technology that creeps into companies with substantial accumulated experience advantages over their competitors.

Where should you locate your facilities?

Location. The issues of size and process technology can have strong location implications. Conventional wisdom in heavy industry suggests that facilities should be located near the source of raw materials (steel, timber, petrochemicals) or near the customer (automobiles, beer, glass containers). A more important consideration in high-technology companies is the source of technical talent and associated technical development activity. One entrepreneur, when asked why he placed his new company in Silicon Valley in California when his customers were predominantly in the southeastern United States, stated that he had selected a location close to his key "raw material": technology.

It is not a coincidence that high-technology enterprises—particularly in electronics and bioengineering—have clustered around major research universities in this country. First, the research universities have provided students, faculty, and facilities that have been key to the early and continuing research in those industries. Second, the clustering itself has been beneficial: professional colleagues share technological ideas and one firm's technology "output" (products or services) may be another's technology "input." The support services, or what the economists might call the infrastructure, for high technology—from subcontracting firms to personnel recruiting consultants to patent attorneys—are more fully developed in areas where high-technology firms have clustered. Entrepreneurial and smaller firms, which need both outside support

Locate near key resources: engineers and technological ideas.

services and trained personnel lured away from larger firms, should locate where they can benefit from this clustering.

Larger, more self-sufficient firms should strongly consider locating new facilities away from these hotbeds of high technology, where engineers' and technicians' salaries are often bid up and venture capitalists eagerly remind key managers and technical personnel of excellent opportunities to start new companies.

I noted earlier the policy of some larger high-technology companies to scatter relatively small facilities to avoid the impersonality of large manufacturing complexes. Other advantages may attend a policy of expanding away from the traditional high-technology areas. Some companies, with Hewlett-Packard again being a prime example, have established criteria for locating new facilities that have much more to do with the projected quality of employees' lives than with the locations of markets, materials, or technology. These criteria focus on such factors as strong public education systems, recreational opportunities, airline service, and availability of continuing education for employees. These criteria lead to sites that resemble the San Francisco Bay Area or the Boston area in amenities, but with lower costs of living and lower facility costs. Recruiting and retaining engineers and capable managers is typically the overriding consideration in location decisions.

Small companies benefit from clustering.

International location of facilities is an increasingly important concern of technology-based companies. Offshore manufacturing sites have been sought in Southeast Asia to take advantage of low labor costs in semiskilled, labor-intensive assembly operations. However, as labor rates in these emerging countries have increased and process technology, particularly in the semiconductor industry, has evolved to lower levels of labor intensity, the cost advantage of offshore manufacture is less compelling.

Location of both manufacturing and engineering facilities in developed countries, particularly Europe and, to a lesser extent, Japan, has become increasingly necessary. In some instances, governments have designated technology-intensive industries as strategically significant and have given them tariff protection or incentives in the form of tax relief, low-inter-

International location is as much a marketing as a production issue.

est loans, or training grants. Effective participation in the markets of such countries may demand that a U.S. manufacturer set up local operations or enter into a joint-venture arrangement with an existing local company. Overseas customers may demand some local presence of their suppliers, particularly if substantial amounts of technical interchange between customer and supplier are anticipated and if after-sale service is a major consideration in the buying decision. Thus, management should recognize that international location is as much or more a marketing issue than a production issue.

∴ Capacity decisions are the key long-term decisions for manufacturing managers. Manufacturing strategy, evolved from and consistent with overall business strategy, is implemented by means of capacity planning. Starting from a solid long-term sales forecast that recognizes the cost-volume-profit relationships inherent in the company's experience curve, managers must decide how much capacity to build and where to build it and evaluate the financial risks and marketing rewards of leading the market with capacity or lagging it. The addition or relocation of facilities provides an opportunity to incorporate new process technology. Viewed over a five-year period, almost any high-technology productive capacity will need substantial design alteration. Managers must remain alert to these needs and opportunities.

Field Service: At the Marketing-Production Boundary

I include the function of field service in a chapter devoted to the boundary between marketing and production because field service operates right at that boundary. Field service personnel are, in a sense, extensions of production because they both install and repair equipment and systems, and extensions of sales because they have extensive interaction with customers.

Producers of technology-based equipment and systems find that a primary concern of their customers is the availability of timely and competent field service. Thus, manufacturers of computers, peripherals, large software systems, analytical instruments, medical instrumentation, process control systems, word processing units, telephone switchboard equipment, duplicating equipment, microcomputer-controlled processing equipment, and many other devices, instruments, and systems (including the associated software) are challenged to provide reliable and responsive field service. The field service must also, of course, be cost-effective.

Field service is an increasingly important competitive weapon.

Tasks

The obvious field service activities extend well beyond simply responding to customers' calls for emergency service. Frequently, the same staff that provides after-sale service is also called upon to install the equipment or system and provide the necessary operator training. Field service engineers are also responsible for effecting warranty repairs. Most companies that maintain field service staffs offer their customers preventive maintenance contracts (which may also provide for some, or all, emergency service coverage).

The not so obvious functions of field service are potentially very rewarding. The field service department might make a profit. I use *might* advisedly because many high-technology companies operate their service departments at a loss, justifying the loss as an additional, but essential, marketing expense. A lack of geographical concentration of installations can greatly complicate the service activity in the early years of a technical company. The operation is likely to be either somewhat overstaffed, so as to provide adequate coverage in all regions, or tolerant of high travel expenses. Losses on service cannot typically be tolerated for long. Management should be particularly alert to the need for corrective action when the loss grows, rather than shrinks, as the base of installed equipment and systems expands. Either service charges or the extent of service provided must be altered. These adjustments can be particularly painful for young companies that, in their early years, underpriced both emergency and preventive maintenance and "overserved" customers in their enthusiasm to ensure their satisfaction.

Consider field service an extension of marketing.

There is, nonetheless, good reason for viewing field service activities as an extension of marketing. Opportunities may arise to upgrade a customer's equipment, sell additional options, or add more units to expand capacity. Field service engineers should accept the responsibility, in concept if not in fact, for selling the customer additional equipment or systems manufactured by their company.

Field service personnel who are properly trained, motivated, and alert can provide valuable intelligence to the field sales force. If field service engineers learn of competitive pressure or a level of customer dissatisfaction that may lead to purchases from a competitor, they should transmit this information to the responsible salesperson. A field service engineer may learn that a prospective customer has visited an existing installation; such information regarding a prospect is also valuable to the sales force and should be passed along.

I spoke in Chapter 2 about marketing's responsibility for conveying new product ideas and suggestions from the marketplace to engineering, bearing in mind that customers originate a sizable percentage of product innovations. Field service engineers are in an even more advantageous position than salespersons to interact with customer personnel, particularly manufacturing and process engineers, and to elicit comments and suggestions that can lead to product improvements, additions of features, and entirely new products. The service organization is also a valuable source of engineering design changes aimed at reducing costs, particularly the costs of warranty repair and of life cycle costs to the customer (including operating costs, preventive maintenance, and emergency repair). Service organizations should compile statistics on the frequency of particular malfunctions and failures of technology products (including software); analyses of these data can reveal design weaknesses of which the engineering staff was previously unaware.

Reporting Relationship

The reporting relationship of field service varies by company. When it reports to marketing, as is typical, the customer is likely to be well served—even excessively served. However, cost-effectiveness suffers. When field service reports to manufacturing, preoccupation with costs often occurs and future sales may be jeopardized by inadequate attention to customer complaints. Management should strongly consider organizing field service to report to the general manager as a co-equal function with marketing, production, and engineering. This structure permits—indeed, requires—the field service manager to make the necessary trade-offs between customer satisfaction and operating costs.

Field service should report to the general manager.

However organized, field service must have extensive interaction with both marketing and production. Field service carries a major responsibility for customer satisfaction. The level of customer satisfaction, in turn, has major implications both for additional orders and for positive word-of-mouth promotion to order prospects. Field service relies upon production for replacement parts, reliable estimates of installation schedules, and a level of quality that minimizes warranty repairs. Again, extensive interfunctional communication characterizes the field service activity, just as it characterizes so many activities in high-technology companies.

Key Challenges for Service Managers

Field service managers in high-technology companies face some unique problems that derive from the accelerating demand for competent service engineers coupled with shortened product life cycles. Five challenges that seem to command a large share of service managers' attention these days are recruiting and retaining personnel, simplifying the process, scheduling, controlling inventory, and formulating policy regarding obsolete equipment.

Recruiting and Retaining Personnel. The recent rapid growth in the demand for field service personnel, particularly as com-

Demand for field service can expand even when sales stop growing.

puters and computer-based systems have spread throughout industry, has caused both a critical shortage of trained and trainable personnel and a sharp escalation in wage rates. No end to these trends is in sight. Field service is a labor-intensive activity. The need for service technicians is a function of both the company's current rate of sales and the size of the installed base of equipment or systems at customer sites. Thus, the demand for service personnel expands even when the growth in sales of new equipment and systems flattens.

Several responses to this shortage of trained and affordable service engineers are possible. The first is competent personnel management. Field technicians typically operate at some physical distance from manufacturing facilities or sales offices; steps must be taken to reduce their psychological distance from the company mainstream. Frequent communication (and not solely by means of the telephone), appropriate tools (including manuals, checklists, and diagnostic aids), and recognition for jobs well done are essential. A field service engineer who feels alone, unassisted, or underappreciated is a ripe target for recruitment by another growing technical firm.

The best of the service engineers must experience continuing challenges and must foresee personal growth opportunities. A number of career paths can be identified, including ones leading to management (local, regional, or home office

Create career paths for field service engineers.

management of the service function, for example) and others requiring technical specialization. A service engineer might be cross-trained on both hardware and software or on a broader set of the manufacturer's product line. He or she may be designated a specialist or resource person, available to other service engineers with particularly knotty problems. Or selected positions in production, sales, or engineering may be identified as advancement opportunities for the particularly competent among the service staff.

Managers should be imaginative in identifying nontraditional sources of candidates for field service positions. Retiring military personnel have proven to be one such source. They have often been well trained technically and are accustomed to a mobile life style. Manufacturing personnel—in final checkout or quality assurance, for example—may also view field service as an interesting new challenge.

Simplifying the Process. The need for field service personnel may be reduced—and recruitment facilitated—if the tasks are simplified. For example, detailed troubleshooting at the customer's site can be reduced by imaginative design of the equipment or system so that modules or subassemblies can be easily replaced. Electronic equipment manufacturers have for years "swapped out" printed-circuit boards rather than troubleshooting them at the level of the individual electronic device on the board. Other manufacturers are now adopting the process with respect to mechanical and electromechanical components as well as electronic components. These steps can lessen the technical sophistication demanded of the field staff and thereby broaden the pool of prospective employees with appropriate skills.

Clever product design and testing procedures can reduce both service time and travel.

Troubleshooting of software can sometimes be accomplished over the telephone; that is, the manufacturer's service group may be able to operate diagnostic software in a customer's computer system, analyzing the results in its own home office system, without visiting the customer's site. Such labor-reducing and travel-reducing steps are becoming increasingly prevalent and necessary.

A more extreme step is to delegate or subcontract the service function to others. Delegation to the customer is the first step to consider. If provided with adequate maintenance manuals and instructions, perhaps backed up by telephone assistance from the manufacturer, the customer may be able and willing to effect repairs or at least to swap the appropriate component or subassembly. If the equipment or system encompasses a major subassembly purchased on an OEM basis from another manufacturer, the maintenance of that subassembly may be contracted to that manufacturer. (However, customers are understandably reluctant to agree to a split in the responsibility for the maintenance of a complex system. Malfunctions frequently occur at the interface between these major elements and finger pointing by the various manufacturers does not lead to an expeditious solution of the customer's problem.)

Try to shift some parts replacement and travel functions to the customer.

The recent advent of retail stores devoted to personal computer and office automation equipment offers another opportunity. The customer may be asked to do some diagnostic work and then bring the malfunctioning element of the system to the retailer for repair or replacement. In effect, the customer, rather than the field service staff, is being asked to do the traveling. The Bell Telephone operating companies have just recently instituted such a change in service policy.

Scheduling. Companies with uneven shipments encounter particular problems in scheduling the field force. I spoke earlier of the end-of-period syndrome (shipments concentrated at the end of the accounting period) that plagues too many technically based companies. This uneven shipment pattern can create havoc for a field force that is expected to install the equipment and provide warranty repairs. The demand for installation follows the uneven shipment pattern, thus imposing very uneven work loads on the field service staff, as well as overstaffing to meet these peak demands. Shipments that are rushed out in an attempt to meet the company's earnings-per-share goals for the accounting period are likely either to be inadequately tested and checked or to be shipped incomplete (notwithstanding the production staff's assurances to the contrary).

Linear shipment schedules pay off in improved field service scheduling and cost control.

When quality suffers, an early indication will be an excessive demand for warranty maintenance. This demand is likely to come in the midst of, or on the heels of, the peak demand for installations. The answer is linear (or even) shipments, with attendant benefits for customers, manufacturing, inventory investment, company profitability, and field service scheduling.

Controlling Spare Parts Inventories. An important cost attendant to providing responsive field service is the cost of inventorying the spare and replacement parts and subassemblies needed to effect rapid repair. (When costs of downtime are extremely high, customers themselves may elect to carry a substantial inventory of spare parts.) Most manufacturers find that they must maintain service parts inventories separate from the production inventories in order that the random demand for spare parts not disrupt manufacturing flows. Investments in these service parts inventories are high and are a major contributor to the slow rate of inventory turnover at many technical companies. (I discuss this further in Chapter 6.)

When setting both service and spare parts prices and evaluating the profit performance of the field service organizations, you should give careful attention to the cost of carrying these inventories—the cost of the capital tied up, the cost of maintaining good control of the physical inventory, and the necessarily high cost of obsolescence of the inventory, particularly as product life cycles shorten. Spare parts should be priced high; gross margins of 75 and 80 percent are not at all uncommon. These high prices prove to be necessary in light of the indirect costs of supplying these spare parts to customers.

Many technical companies regularly reanalyze the question of whether they should maintain spare parts inventories regionally or centrally. The number and completeness of spare parts depots required are a function of (1) customers' tolerance regarding downtime, (2) the cost and availability of air freight to customers' sites, (3) the predictability of demand for particular parts or subassemblies, and (4) the required investment in the parts to be inventoried. Most manufacturers should pursue the middle course, maintaining redundant inventories of the most frequently demanded and least costly parts while relying on air freight to provide timely delivery of the more expensive and less frequently demanded parts and subassemblies. Unfortunately, this statement begs the question of just *which* parts should be inventoried regionally. A study within the context of your particular company of the four factors just listed can yield a set of guidelines for making these part-by-part decisions.

Recognize the cost of inventorying spare parts and price them accordingly.

Formulating Policy on Obsolete Equipment. How long should a manufacturer commit to its customers to provide service capability and spare parts for a particular product line or model? Forever is too long. The company needs to develop and enunciate a policy regarding its commitment to obsolete equipment. The policy should be a function of (1) the rate of technology and product change in the industry, (2) the company's image with its customers regarding service, (3) its ability to cause customers to purchase new models to replace obsolete ones, and (4) competitive pressures.

Let your customers know your policy on obsolete equipment and stick to it.

Premature abandonment of the service commitment to a product can have serious customer repercussions, but an excessively long commitment will do the following:

1. Increase the cost of providing service because new members of the field force will have to be trained to repair the obsolete equipment
2. Balloon inventories of spare parts as spares for old equipment turn over more slowly
3. Increase inventory write-offs as demand for particular parts approaches zero
4. Eliminate the opportunity to sell new, updated models to customers to replace the obsolete equipment

A wide range of policies typically proves acceptable to customers, but customers do not like surprises. It is imperative to be explicit about the policy, announce it early, and stick to it.

∴ Field service engineers must adopt both a production and a marketing viewpoint, regardless of whom they report to. Besides their obvious tasks, field service personnel can provide both important sales intelligence and useful feedback for product improvements (particularly to achieve higher quality and reliability) and for new products. Field service managers face particularly thorny people management issues that arise because of the nature of the task and the shortage of qualified personnel. Cost control in the service area demands careful scheduling, active management of inventories, and an explicit policy on obsolete equipment.

Highlights

- Tensions abound at the marketing-production interface; these tensions relate particularly to reconciling the order forecast and the production build plan, making and keeping delivery commitments, and managing the order-receipt-and-delivery cycle on individual customer orders.

- Good short-term forecasting, imaginative framing of the build plan, timely freezing of schedules, clear procedures for making delivery commitments, using blanket orders, and achieving even shipment flow can lessen these tensions.

- Manufacturing defines its strategy by ranking cost, quality, flexibility, and dependability.

- Changes in priority among these four objectives must be anticipated as the company's products and markets mature.

- Capacity planning involves the key decisions that implement the manufacturing strategy and is the mutual responsibility of marketing and production.

- The learning or experience curve, linking costs, volume, and prices, affords competitive threats as well as opportunities.

- The key capacity planning issues are size, timing, design, and location.

- Size is influenced by issues of coordination and the opportunity for focus, as well as by conventional economies of scale.

- Timing must balance the financial risks of leading the market with the marketing risks of lagging it.

- Capacity design must incorporate process technology that is consistent with the maturity of the products and markets and with the company's experience curve.

- Capacity location decisions turn on personnel policy, technology source, and marketing issues.

- Field service engineers operate at the boundary of marketing and production and must view their jobs as extensions of both functions.

Notes

1. See, for example, Robert H. Hayes, "Why Japanese Factories Work," and Stephen C. Wheelwright, "Japan—Where Operations Really are Strategic," *Harvard Business Review* (July–August 1981), pp. 57–66 and 67–74.

2. See the PIMS studies (Profit Impact of Market Strategies) compiled by the Strategic Planning Institute, 955 Massachusetts Avenue, Cambridge, Massachusetts 02139 (telephone number 617-491-9200).

3. Wickham Skinner, "The Focused Factory," *Harvard Business Review* (May–June 1974), p. 114.

Chapter Four

Coordinating Production and Engineering

This chapter covers the following topics:

- Determining the Company's Technical Policy: Key Management Decisions
- Communication Between Engineering and Production: A Critical Factor
- The Optimum Degree of Vertical Integration for a High-Technology Company

The production staff laments that, if only the design engineers would pay more attention to the producibility of their designs, manufacturing's job would be much easier. And, just when the production process is becoming routinized, why must engineering change the design? In turn, engineering staffs indict manufacturing for inadequately skilled employees. After all, prototype units of the new product were built successfully in engineering's model shop; so full-scale production should be a simple matter of scaling up.

Chapters 2 and 3 covered the management issues that arise at the intersections of marketing and engineering and marketing and production. Another traditional battleground in high-technology companies is the interface between production and engineering. Orchestrating the potentially conflicting views and priorities of engineering and production is a key role of the high-technology general manager.

Both the production and the engineering functions must be deeply and critically involved in the firm's technology. Too many companies view technology as being related solely to products, and thus entirely the province of the design engineers. Increasingly, the successful high-technology companies appreciate the importance of process—as well as product—technology. Thus, the production department must also view technology as a source of strategic advantage. Both functions should participate in formulating as well as implementing what I refer to as the firm's *technical policy.*

I begin by focusing on the interrelationship of product and process technology. Here we find the twin, but often competing, goals of designing to cost and designing to performance. We consider the optimum mix of product development (design) engineering and process (manufacturing) engineering.

Much of the traditional tension between engineering and production turns on the problem of communication—with respect to new products, existing products, and changes to both. What are the respective responsibilities of the two functions? What formal, documented procedures are appropriate?

Although all functions of the high-technology company are concerned with the strategic issue of vertical integration, the degree to which the company backward integrates is determined largely by the advantages to be gained by the company's achieving self-sufficiency in certain key product or process technologies. Thus, again, the decision on backward integration must be made mutually by engineering and production and must be consistent with the firm's technical policy.

In exploring the boundary between production and engineering, I focus on the key decisions that define the company's technical policy; the product-process matrix discussed in Chapter 3, now with attention to its implications for the proper mix of engineering capabilities; engineering documentation, a key means of communication between the two functions, particularly as it relates to new products and changes to existing products; and the issue of vertical integration— how a high-technology company decides when and where to be self-sufficient and when and where to rely on vendors.

Determining the Company's Technical Policy: Key Management Decisions

Process technology may be as key to your competitive position as product technology.

The high-technology firm's technical capabilities are centered in both the production and the engineering functions. Managers of firms with important foundations in technology must guard against thinking of all technology as product based; process technology is often key to the firm's competitive positioning.

Answers to the following questions define the role of technology within your firm:

1. What technologies should the company invest in? How broad or narrow should the set of technologies be?

2. What should be the relative mix of product and process technologies?

3. How close to the state of the art should the firm be? (Refer again to the research-engineering continuum in Figure 2-1.) How proficient should the company become in both understanding and applying the technologies?

4. What sources of technology should be relied upon? How integrated should the company be in technology? Should external sources, including contract research and suppliers, be relied upon?

5. Should the company lead or lag in the application of new technologies? Do the benefits outweigh the risks of being first?

Be explicit in defining your technical policy.

We shall see in this chapter that the answers to these questions both influence and are influenced by the linkages between products and processes and the degree of the firm's vertical integration. Explicit agreement on technical policy, particularly regarding the second and third questions in the previous list, helps define the appropriate communication system between engineering and production. Finally, the answers to these questions have important implications for both the company's financial structure and its organizational policies. I discuss these implications in Chapters 7 and 8.

◈ *Products and Processes: An Inextricable Link*

A product's design influences the methods used in its manufacture. Ideally, process technology—manufacturing techniques and capabilities—should in turn influence product design. Thus, product technology and process technology are inextricably linked.

Engineering is crucial to achieving manufacturing's cost, quality, flexibility, and dependability objectives.

In Chapter 3, I discussed the four objectives whose ranking defines the company's manufacturing strategy: cost, quality, dependability, and flexibility. Engineering plays a critical role in attaining these objectives. Some engineering designs are inherently more reliable or achieve tighter specifications (provide higher quality) than others. Such designs, however, may necessitate some sacrifice in product cost or in manufacturing flexibility. For example, injection molded or die cast parts may permit both higher quality and lower cost than a machined part, but the resulting restrictions on design changes represent a loss of flexibility. Electronic instruments can be designed for assembly by skilled technicians—at high cost but great flexibility—or they can be designed at a somewhat greater investment in engineering time and tooling for assembly by semiskilled personnel—at reduced manufacturing cost and perhaps with some loss in flexibility.

I argued in Chapter 2 that the management of a high-technology company should make a conscious decision regarding where on the product-process matrix its various product lines fit (see Figure 3-3). Moreover, it needs to recognize that, as technologies mature and market and competitive pressures change, this position may evolve, typically down and to the right toward higher volumes and more automated manufacturing. As this evolution occurs, the changes required in both the organization of manufacturing and the production techniques employed are reasonably obvious. What is not so obvious, but every bit as important, is that the engineering skills and organization also need to evolve.

Evolution of Engineering Skills

Be sure engineering skills and organization evolve along with those of production.

Early in the life of distinct technologies and product lines, engineering is state of the art and emphasis is on understanding the new technology and harnessing it to the particular product-market segment. Many changes and refinements in the products occur, as early testing and customer evaluation reveal design shortcomings or opportunities for improved performance. The link between engineering and marketing is strong, and manufacturing is relied upon to produce, albeit in small quantities at relatively high cost and with skilled labor, what engineering has designed.

As the products and technologies mature, engineering's focus must necessarily turn more to the questions of producibility, reliability, and cost. Now the links between engineering and production must strengthen, and engineering's emphasis moves somewhat away from science and toward manufacturing engineering. Design changes and improvements derive from manufacturing considerations, as well as from marketing or customer considerations. Improved tooling or automating certain processes often require changes in design, changes that are transparent to the customer.

The engineering staff that was appropriate during the early stages of the life of both the products and the technologies may be inappropriate—or at least less effective—later in those lives. The engineer who is excited by the frontiers of new technology will frequently be impatient with and frustrated by demands for product redesign to improve manufacturability. Similarly, the engineer knowledgeable about sophisticated manufacturing processes and excited by elegant design for high-volume manufacturing will be impatient with and frustrated by the incomplete and fluid designs that characterize new products that incorporate new technologies.

Design to Performance Versus Design to Cost

Another way to view this shift in technological emphasis and engineering activity is to consider the relative importance of performance and cost in the overall design criteria. Is the company designing to performance, with cost as a subordinate objective, or must cost reduction be uppermost in the designer's mind? In certain product-market segments, performance considerations dominate price. In other segments, the reverse is true. Where performance dominates, the market can be said to be performance elastic—that is, improved performance can bring forth substantial additional buyer interest. Where price dominates, the market is classically price elastic—reductions in price bring forth substantial new demand.

Are your products performance elastic or price elastic?

A firm serving performance-elastic markets must emphasize state-of-the-art products in its technological policy as well as leadership in incorporating new technology. It must rely on many alternative sources of technology and emphasize engineering innovation, incorporation of new technology, and maintenance of a performance edge over competition. Customers eagerly seek the latest in technology and the improved performance made possible by the rapidly evolving technology. Prices in such markets are high and volumes are typically low. Manufacturing costs, although high in an absolute sense, are low as a percentage of price. Manufacturing engineering can and should be deemphasized; the company's attention and resources must be focused on new technology that can lead to improved performance.

High-volume markets quickly become price elastic.

Markets for high-technology products have traditionally been perceived as performance elastic. Increasingly, however—and particularly in consumer electronics—high-technology products are competing in price-elastic markets. A number of years ago, the markets for handheld calculators and digital watches moved from being performance elastic to being price elastic. More recently, home computers, video games, and video recorders have followed the same pattern. As these shifts have occurred, the successful companies in these markets have shifted their technological policies away from the state of the art and toward proficiency in implementing the technologies. Design engineering focuses somewhat less on designing to performance and somewhat more on designing to cost. Process engineering takes on increased importance.

Probably no U.S. technology-based company has been more explicit or effective in this design-to-cost strategy than Texas Instruments (TI). Known for pursuing aggressive pricing policies based on experience curve economics, TI treats cost projections as central to both its product design parameters and its business strategy. While acknowledging the importance of high volumes, that is, accumulated experience, a top executive at TI states:

At TI, designing to cost is part of the culture.

> High volumes do not drive costs down; they merely provide the opportunity. At TI, capitalizing on this opportunity starts with "Design to Cost." This involves deciding today what the selling price and performance of a given product must be years in the future and designing the product *and the equipment for producing it* [emphasis added] to meet both cost and performance goals. Stated another way, unit cost is a primary design parameter. It is a specification equal in importance to functional performance, quality, and service. This parameter takes the

form of a timetable of steadily decreasing costs over the entire lifetime of the product. . . .

Since the founding of TI, we have worked to create an organization and an institutional "culture" in which continuing productivity increases, cost reduction, and the Design-to-Cost approach are viewed as moral obligations to society.[2]

Manufacturing Engineering Department

Still another view of this shift in technological emphasis is to consider just what role the manufacturing engineering department should play. Manufacturing engineering must be strongly emphasized in a design-to-cost company. At Texas Instruments, manufacturing engineering plays a key role in achieving the company's objectives of designing to cost.

Virtually all high-technology manufacturing companies maintain a department bearing the title of Manufacturing Engineering or Process Engineering. As the name implies, its functions are central to this discussion of the boundary between engineering and manufacturing. These engineers are the key link between the development laboratory and the production floor.

Manufacturing engineers are the key link between the lab and the factory.

Where process technology is accorded high priority in the firm's overall technological policy, these engineers should be given substantial authority and responsibility for influencing product design to improve its manufacturability and improving process design, often including substantial investments in equipment and tooling. In addition, manufacturing engineers provide technical support to line manufacturing personnel, lending technical assistance to and troubleshooting on the production floor. In short, manufacturing engineering is essential to the twin objectives of low cost and high quality.

The Japanese electronics companies excel at manufacturing engineering. The argument is often made that research and development staffs in Japanese companies are not particularly innovative, that they rely on improving Western products that they have "reverse engineered." (If this argument was ever valid, it is almost certainly becoming less so.) Japanese engineers have focused their attention on the weakness in many U.S. high-technology firms: manufacturing engineering. By emphasizing producible designs, flexible and high-quality tooling, and extensive communication between the engineering and production personnel, Japanese companies have realized startling economies in product cost with improved product quality—and yet retained great production flexibility. These results have given the Japanese enormous advantages in the marketplace. They have gained those advantages in a straightforward manner: through diligence and attention to detail.

Don't relegate your manufacturing engineers to second-class citizenship.

By contrast, in this country, manufacturing engineers have for many years been considered second-class citizens of the engineering world. The truly capable engineers were assumed to be in design, and individuals who really understood manufacturing were engaged in production supervision, where the "action" is. Indeed, many companies assigned to their manufacturing engineering department those design engineers who were not particularly effective in the laboratory and those production supervisors who did not quite measure up. The Manufacturing Engineering Department often became the dumping ground, and the department developed its reputation accordingly.

The Japanese have taught us that we ignore manufacturing engineering at our peril. Except in those state-of-the-art companies where performance considerations thoroughly dominate cost considerations, high-technology companies should emphasize manufacturing engineering, assigning their best engineers to that function, and recognize that market acceptance, and thus company health and profitability, turn every bit as importantly on cost (price) and quality as on spectacular product technology.

Should manufacturing engineers report to manufacturing or to engineering?

Where should manufacturing engineering report in the organization—to manufacturing or to engineering? Although the reporting responsibility is typically to manufacturing, opinion is understandably divided. Early in the company's life, when technology and products are new, the Manufacturing Engineering Department should report to engineering because its important role then is to influence design rather than to automate manufacturing. Indeed, in very small, entrepreneurial, technical companies, the manufacturing engineering activities—particularly attention to producibility— are often simply a part of the design engineers' role, with no separately designated department. The small firm enjoys intense communication and interaction; so organizational formalities are unnecessary.

As the company, technology, and products mature, the Manufacturing Engineering Department's focus should shift more to production automation—taking steps to steepen the slope of the learning curve—and attaching the department to the manufacturing function is more appropriate.

A periodic change in reporting relationship can be healthy, particularly if the company cycles between emphasizing new products and emphasizing improvements in existing products. Such shifts can strengthen the Manufacturing Engineering Department's liaison role, assuring that neither engineering nor manufacturing viewpoints become dominant. A periodic shift in reporting relationship represents a kind of job rotation for the engineers. I stress in Chapter 8 the benefits to be gained by job rotation in a high-technology company.

Consider changing reporting relationships periodically.

∴ Every high-technology company must develop an explicit technical policy as a basis for formulating both engineering and manufacturing strategies. Company managers should define the policy in terms of the breadth and depth of its technical competence and the sources and timing of the introduction of new technology. This policy dictates the links between product and process technologies, establishes the priority of designing to performance or designing to cost, and implies the relative roles of design engineering and manufacturing engineering in achieving competitive advantages.

Communication Between Engineering and Production: A Critical Factor

Extensive communication between engineering and production is critical to implementing the firm's technical policy. Communication must be both formal and informal. Informal

Encourage both informal and formal communication.

communication should be encouraged in all the ways that have become common in high-technology companies, from "beer busts" to technical symposia to off-site, multiday discussion sessions.

Regardless of the amount of informal communication, formal communication is also essential. The formal communication system between engineering and production must deal with three important, related, but distinct, challenges:

1. Introducing new products from the development laboratory to the production floor
2. Providing the optimum—neither maximum nor minimum—level of documentation on existing products
3. Facilitating orderly and cost-effective changes to products now in production

Introducing New Products to Manufacturing

Payoffs accrue to the company that moves new products from engineering to production smoothly.

Handing over the new product from engineering to manufacturing tests the cooperation and communication between engineering and production personnel as does no other activity. The high-technology company that manages this transition well stands to gain timing, cost, and quality advantages that can have substantial payoffs in the marketplace.

Where departmental barriers are high and engineers are encouraged or permitted to be myopic, design engineers will attempt to maximize product performance and manufacturing engineers will try to redesign the product to reduce its cost. Such a two-step process is highly inefficient and very time consuming. Optimizing across the conflicting priorities of cost and performance—often called value engineering—must be the responsibility of everybody engaged in new product creation, including marketing managers who have a hand in setting the new product's target specifications.

Extensive communication must both precede and follow the formal transition from engineering to production. Periodic product design meetings that involve design engineers, manufacturing engineers, and material planners (and sometimes product managers from marketing and others as well) should be held monthly during the early design stage and perhaps weekly just prior to and following the "hand over" of the new product from engineering to production. In these meetings, the design staff consults with manufacturing personnel regarding design alternatives under consideration and gains insight into the issues of producibility as the product or system is being designed. Realistic tolerances are specified, and engineers are encouraged not to tighten tolerances to expedite design or to gain an additional margin of safety. (Excessively tight tolerances almost always increase manufacturing costs.) Manufacturing should share with the design team its experience with present vendors and subcontractors as engineering is selecting sources for parts or processing for the new product or system.

Hold periodic product design meetings before, during, and after the transfer.

Both manufacturing and engineering personnel must be aware that the design process is generally not complete when manufacturing commences. Design errors that need attention may be uncovered; change requests initiated by manufacturing to facilitate fabrication or assembly must be evaluated; and operating performance that met specifications in the laboratory but cannot be replicated on the manufacturing floor must be reassessed. Proper reliance on prototype units and pilot production runs before full-scale production is attempted can reduce costly errors.

Prototype, Pilot, and Production Runs. Engineering typically produces prototypes (the first one or two units of a new product or system). The engineers and design technicians construct them at considerable cost, frequently building and rebuilding them. They use techniques appropriate to the lab but inappropriate to full-scale production—"breadboards" instead of printed-circuit boards and fabricated instead of cast metal or injection-molded parts—in order to facilitate design changes and minimize tooling costs and lead times. These prototypes, which are often necessarily quite different physically from the units ultimately supplied to customers, should be both thoroughly tested in the laboratory and subjected to some field testing. The purpose of prototypes is to prove design concepts and confirm product specifications.

Once the basic design concepts have been proven in prototype and satisfactory operating specifications have been met, a pilot production run (the production of a limited quantity) should be initiated. The design used for the pilot production run should be the one that is expected to be used in full-scale production—for example, breadboards are now replaced with printed-circuit boards, and substantially more investment is made in tooling. The purpose of the pilot production run is to test product producibility and to work out any bugs in the final design before the company scales up to full production. (When total anticipated volume of the product or system is small, this pilot production step can be eliminated.)

Prototypes prove design concepts and pilot runs prove producibility.

In some companies, pilot production runs are undertaken by the engineering department and, in others, by the manufacturing department. The exact reporting relationship is not particularly significant. What is important is to recognize that pilot production runs are inevitably the joint responsibility of engineering and production.

Freezing Designs. Before full-scale production is undertaken, the design must be frozen, after which time formal engineering change notices are the only mechanism for effecting changes. In the absence of a pilot production run (and sometimes even with it), the point at which design becomes final, or frozen, is often unclear. At the prototype stage, the design must be allowed to remain fluid, permitting design changes at minimum cost and documentation. But it is human nature to seek almost endlessly for small improvements and refinements. This propensity is as true for the design engineer who is a parent of the new product as it is for the artist in her sculpture or the writer in his manuscript. Just as editorial changes are expensive to effect once the manuscript has been set in type, so product design changes are expensive to effect once manufacturing has commenced.

Manufacturing and engineering must agree at what point the design is frozen.

Thus, at some point, the design of the new product must be frozen, and both manufacturing and engineering must agree upon that point. Subsequent design changes can no longer be made unilaterally by the design engineer, as they could during the prototype phase.

You can often gain important timing and cost advantages by freezing certain portions of the design before other portions. For example, in a complex computer-based system, the selection of the system's minicomputer can and should be frozen long before other portions of the system are designed, in order to provide sufficient time for the programmers to develop the necessary software and for purchasing to negotiate OEM contracts with the minicomputer supplier. Sequential freezing is appropriate: Freeze parts or components known to have long procurement lead times early; leave standard components and those parts requiring little or no tooling unfrozen until late in the design cycle. A complex design project can usefully be subjected to PERT (program evaluation and review technique) analysis to reveal the critical design concepts or components that need early freezing.[3] This process of sequential freezing of portions of the design implies close working relationships and much communication throughout the engineering organization and between engineering and production.

Sequential freezing saves money but requires close communication within engineering and between it and production.

Top managers of high-technology companies should see to it that procedures for freezing designs are both established and adhered to.

Using a "Skunk Works." Some technology-based companies have successfully used an unusual organizational technique to expedite new product design and introduction. When a new product requires (1) a number of engineering disciplines, (2) careful attention to manufacturability and cost, and (3) a telescoping of the design and introduction stages, a separate task force may be created, drawing personnel from a number of functional departments in the company. When this task force is assigned separate facilities, sometimes with extra security against industrial espionage, these facilities are often referred to as the "skunk works."

The objective is to recapture the advantage of the small company: high motivation, focused purpose on a single product, system, or process, and intensive and informal communication with a minimum of organizational barriers. The task force is accorded (or assumes) high prestige in the organi-

Consider a skunk works to recapture the advantages of a small firm.

zation, and assignment to it is eagerly sought. Extra resources are typically made available to the task force.

Arguing against the establishment of a skunk works is the fact that creating one or a series of these task forces can be disruptive to the organization. Other development projects may be interrupted and key technical personnel assigned to the task force may be unavailable for informal counsel and advice on projects to which they are not formally assigned. Acceleration of the design can also cause some loss of efficiency.

Use the skunk works only when you need fast action.

This organizational device should be used only when competitive conditions demand fast action, either to protect an existing market position or to gain a jump on anticipated competition. Although the device has proved highly effective in a number of instances, resulting in a dramatic product unveiling that left both customers and competitors in awe, its overuse reduces the opportunities for specialization, economies of scale (experience curve economies), and routinizing of procedures.

A variation on this organizational device is to assign a group of engineers to "follow" a new design through the laboratory and onto the production floor. That is, rather than turning over its design (and prototypes) to manufacturing engineering, a portion or all of the engineering design team is assigned the responsibility for moving with the product from the design engineering organization to the manufacturing engineering organization. The trade-off is that the "following" engineers will know the new product in detail, thus eliminating the need for manufacturing engineers to learn the new product, but will be less experienced and probably less capable in attacking the problems of producibility and tooling. However, a design engineer who has spent some time wrestling with new products from a manufacturing engineering viewpoint will be a more effective design engineer when he or she returns to the laboratory and another new product. Again, a type of job rotation has occurred.

Managing multiple products is much more complex than managing just one.

Moving from Single to Multiple Products. Many emerging technical companies—that is, small but rapidly growing companies—encounter real turmoil as they move from relying on a single product to offering multiple products to the market. A small technology-based business focusing on designing and manufacturing a single product is often wonderfully efficient. It minimizes conflicting priorities because all hands are devoted to the single product. As the business grows and more product lines are added to the company's portfolio, choices must be made. The general management task suddenly becomes much more complex.

In engineering, the task of product maintenance engineering on the older products competes for attention with new product development. The need for standardization of components and subassemblies across product lines becomes evident. Compromises between standardization and optimum price/performance suddenly become necessary.

Product maintenance conflicts with new product design.

In manufacturing, quality problems that remained under control because of the undivided attention of manufacturing and engineering on the single product line now drift out of control as technical attention is diffused across many products. The existence of multiple products on the manufacturing floor complicates production scheduling. These products require both unique and common skills and often incorporate common parts or subassemblies that ought to be produced in larger lots.

Shared inventory and common parts complicate manufacturing.

The interaction between engineering and manufacturing was extensive on the company's first product line. This intimate, one-on-one communication needs to be continued on the newer products, but coordination on older products needs to be more routinized.

A key test for an emerging high-technology company is its ability to move successfully from engineering and producing a single product to engineering and producing a portfolio of product lines. The transition requires that a manufacturing engineering function be established, as well as a data base and reference system to aid in standardizing components. Task assignments in engineering must clearly recognize the dual responsibility of product maintenance and product development. A formal documentation and engineering change request system must replace the informal communication that sufficed when the company was small and produced only one product.

Engineering Documentation

Formal communication between engineering and production demands product and process documentation: drawings, bills of material (parts lists), schematics, assembly prints, software listings, and many other elements of paperwork (and now, increasingly, microfiche, computer data bases, videotapes, and other media). Most of these communication media are created by engineering and represent the detailed specification of the product to be produced or the process to be operated.

How detailed should your documentation be?

Level of Detail.   A persistent dilemma facing management in high-technology companies is the decision of just how much detail to incorporate into the documentation of particular products and processes. Detailed documentation, taking the form of prints, parts lists, assembly drawing, process and assembly instructions, and sometimes audio, video, and other nonprinted media, is expensive to create, control, and update. However, skimpy documentation may be risky, allowing design changes to be effected without thorough review. Such incomplete documentation may also inhibit accurate and complete communication among the functional departments of the business.

The dilemma is resolved primarily on the basis of the relative importance the high-technology company places on manufacturing flexibility and product costs. Very detailed documentation is required when (1) production volumes are high, (2) automation and tooling are relied upon to reduce costs and (3) less-skilled manufacturing labor is to be utilized. More elaborate documentation is a prerequisite to the aggressive pursuit of learning curve economies. Such elaborate documentation is not justified, however, when volumes are small, a skilled work force can be relied upon to operate with limited instructions, and design changes are implemented at a rapid

High volumes and automated manufacturing demand detailed documentation.

rate. As a general rule, more documentation is appropriate, justified, and necessary as one moves along the continuum from custom to standard products.

High-technology companies most frequently err on the side of too little documentation. This tendency is not surprising. In the early stages of the life of products and technologies, a minimum of documentation is appropriate. As the company, products, and technologies mature, there is a reluctance to invest engineering time and attention in paperwork on existing products rather than in designing new products. Companies that neglect documentation, however, find they are forever running to catch up with the required documentation.

Don't underdocument.

General managers must strike the proper balance between too much documentation and too little. Despite protests to the contrary from most manufacturing managers, more complete and thorough doumentation is not always appropriate. The right balance is a function of the overall business strategy and of the position of the particular product or product family within the product-process matrix. When the strategy is geared to a succession of new, high-technology products, skimpy documentation is both appropriate and cost-effective. When the strategy depends upon achieving learning curve economies— the company is operating down and to the right on the product-process matrix—complete, up-to-date, and reliable documentation is essential.

Effects on Inventory Control.   Effective inventory planning and control requires very accurate bills of material (that is, listings of individual parts, components, and subassemblies that go into a finished product). Inaccurate or incomplete bills of material preclude using sophisticated planning techniques, such as MRP. The result is that excessive raw material inventories are held in order to guard against shortages. Moreover, the omission of one or more parts from a bill of materials can cause a halt in the assembly process while the missing part is located. The result is that in-process inventories also balloon. Thus, improved inventory control in high-technology companies, an objective stressed repeatedly throughout this book, requires the active participation of engineering, as well as of the production and finance departments.

Accurate documentation is a prerequisite for good inventory control.

Processing Engineering Changes on Existing Products

Life in a technology-based business would be substantially simplified if all documentation, once created, could be relied upon to be both accurate and stable. Neither condition is eas-

Engineering changes require accurate documentation.

ily achieved when both technology and product change is an ever-present fact of life. All engineering changes, whether to improve performance or to reduce costs, must be reflected in changed documentation. In addition, design errors uncovered by engineering or manufacturing personnel (and sometimes by field service personnel) must be corrected and the corrections incorporated into the documentation system.

Thus, requests for changes to existing products can—and should—emanate from all corners of the organization:

Encourage engineering change requests from all over your firm.

1. From engineering to take advantage of new technology or to incorporate a new product feature
2. From the field service organization to improve reliability or to facilitate field repair
3. From purchasing to take advantage of a new supplier or a lower price of a substitute component
4. From marketing to improve the competitive posture of the product
5. From manufacturing engineering to permit the use of more sophisticated tooling
6. From production and inventory planning to permit standardization of components across product lines
7. From production supervisors to reduce tolerances, and thereby costs, or to facilitate processing or assembly in some other way

Changes that reduce costs are an important element of the experience curve phenomenon discussed in Chapter 3.

Just as requests for changes can emanate from all corners of the organization, so implemented changes affect all corners of the organization, including particularly purchasing, inventory control, marketing, field service, production supervision, and cost accounting. Because these organizational units will be affected by the change, they must have a hand in deciding whether the requested change should be adopted (and when), and they must be notified in a timely fashion of approved changes.

The number of change requests may be very high—in the tens for a simple product, the hundreds for a complex instrument, and the thousands for a comprehensive system. Each change is likely to have a ripple or domino effect on documentation. For example, the change of a single component may require a change in the drawing on which it first appears, on one or more bills of materials, on drawings of parts or assemblies farther up the product tree, on assembly instruction sheets, and so forth. Each change may have both obvious and not-so-obvious consequences; these need to be anticipated, evaluated, and, if appropriate, tested.

Anticipate, evaluate, and test the subtle consequences of requested changes.

Technical companies should develop and institute formal procedures and paper flow systems to be certain that all necessary documents are changed as required, that changes do not become incorporated into the documentation before they are appropriately authorized, and that all affected individuals and groups within the organization are aware of the nature and effective date of the change in sufficient time to adapt accordingly. The process must be both rapid and thorough, but it also must be routine, if production and engineering activities are not to grind to a halt either as a result of a preoccupation with processing changes or a lack of coordination among the changes themselves.

Discipline must be built into the engineering change request system so that procedures are not short-circuited. If control of documentation is lost, the following conditions can occur:

Inadequate documentation control leads to quality, inventory, and cost problems.

1. Quality problems multiply as exact specifications of components become impossible to trace and unanticipated consequences arise from unauthorized design changes.

2. Inventory investments and write-offs increase as parts are made obsolete without notice and the production cycle lengthens because newly specified parts are not planned and acquired in a timely manner.

3. Manufacturing labor costs escalate as expediting, troubleshooting, and additional setups consume both direct and indirect labor hours.

Proper handling of engineering changes is the bugaboo of documentation methods in many high-technology companies.

Cost and Benefit Trade-Offs

All changes have both benefits and costs, even those that simply correct drawing errors. The challenge is to make the proper trade-off. Engineering changes that alter the physical specifications of particular components may render obsolete present components now in inventory and necessitate rescheduling of manufacturing work orders or purchase orders with vendors. Such obsolescence and rescheduling costs must be weighed against the advantages to be achieved from the change to decide both if and when the change should be effected. The optimum decision is often to delay the change until present inventories are depleted, until new vendors can be brought on stream, or until other conditions occur that will minimize disruption.

Some changes—for example, in computer software—must be expedited to fix a bug in a particular program, with notification rushed to various parts of the organization and to customers. Other changes in the software—changes designed to enhance capabilities or improve execution efficiencies—should be saved up and incorporated with other alterations in periodic rereleases of entirely new generations of software. Changing software documentation is expensive, and such changes typically require changes in operating, training, and service manuals as well. Batching changes may be efficient, but this advantage must be weighed against the disadvantage of delaying the introduction of an improved product to the marketplace.

Should you batch changes or implement them individually?

The initiator of a change request may be unaware of the full ramifications of the proposed change. A change in part M may require an adaptation of part P or assembly T, expensive reworking of tooling, or a change in maintenance procedure that must be communicated to customers and the field service force. The possibility that the benefit sought from the engineering change request could be more expeditiously accomplished by an alternative change must be evaluated. For example, a problem that could be corrected by a hardware change might also be correctable by a less-expensive software change.

Evaluate all changes on the basis of costs and benefits. Making the trade-offs between the costs and the benefits of change is complicated within most high-technology companies by the fact that the relevant data on both costs and benefits are not readily available to the decision maker. Manufacturing cost penalties or savings may be ascertainable (although even here most cost accounting systems do not reveal the incremental costs), but the tangible and intangible benefits or costs associated with changes in competitive position, in ease of field maintenance, or in vendor relationships are often uncertain. The costs of effecting the change—engineering time, clerical effort on documentation, renegotiation by purchasing, and possible scrapping of inventory—must be factored into the decision.

Engineering change requests must be routed for approval through each affected department: design engineering, manufacturing engineering, material planning, quality assurance, and field service. (In some companies still other departments should formally approve changes.) Each of the evaluators must be alert to the possible need to solicit input from other functions, such as marketing or finance. Checklists and rules of thumb may help streamline the process. An engineering change committee, which is responsible for making the final cost-benefit trade-off when disputes arise, should be constituted.

Require approval of all change requests from all affected departments.

New Models Versus Incremental Changes

I spoke earlier of the importance of freezing new product design, and now I have suggested that engineering changes may occur in large numbers. What factors should management consider in deciding how much product evolution to permit through the engineering change request mechanism?

First, saving up (or batching) engineering change requests in order to effect many changes at one time can have distinct advantages in reducing implementation costs. Disruptions in both production and engineering are minimized.

More important advantages often attend the introduction of a brand new model or line of a product. First, the company's image in the marketplace may be enhanced when it introduces a new product or model that delivers significantly improved performance. The opportunity may exist to leapfrog the competition. A series of incremental changes may not have

Reduce implementation costs and minimize production and engineering disruptions by batching change requests.

the same marketing impact on customers as the introduction of a new product, and competitors may be better able to react to, and sell against, a series of small improvements. When these conditions are present—as they usually are—management should restrict product evolution through small, incremental changes, even when such changes would result in some improvement in performance.

The engineering staff may benefit from an opportunity to start over on a product line, to incorporate new technology or new design concepts that cannot be utilized given the constraints of the present product. Such starting over is, of course, expensive, but new competitors entering the market are not constrained by present products. Thus, if the removal of such constraints represents an important design advantage, management should be certain that its own design engineers are not denied that advantage. For example, the full benefits of a new software language probably cannot be realized without starting over, and the maximum benefit of VLSI circuits is not realized by designing incrementally from present products.

Consider starting over on a product's design rather than improving it incrementally.

Relatedly, new models or product lines, rather than incrementally improved present products, often permit adoption of manufacturing techniques that provide the company with significant cost and quality advantages. The use of robots in fabrication and assembly typically requires some product redesign to make optimum use of the robots' capabilities. The opportunities for automation may not be evident or, if evident, may not be economically justified if product design is accepted as a given. The concept of the product-process matrix introduced in Chapter 3 presumes that both product design and process design are subject to changes and that the changes can and should be related.

The case should not be overstated, however. A market leader, such as IBM in mainframe computers, may need to pay particular attention to thwarting competitors' attempts to copy (often referred to as "reverse engineer") its products. A continuing series of well-planned design changes can severely complicate the process of reverse engineering and permit the leader to sustain a technological and performance edge over its competitors.

Allocation of Engineering Resources

Related to this problem of new products versus incremental changes is the inherent risk in high-technology companies that excessive engineering resources will be diverted from the truly new product to service the existing products. New products are the lifeblood of such companies; the more the company relies on technology to differentiate itself from competitors, the more this statement holds. Two sources of diversion are prevalent: product line maintenance and customer "specials."

Don't squander engineering resources on product line maintenance.

In this chapter, I have been emphasizing that continuing engineering of existing products is not unimportant, particularly as production seeks to improve product manufacturability and reduce its costs and as the need for improved documentation is realized. But such maintenance must not be permitted to consume all engineering resources.

In Chapter 2, I emphasized that customers' requests for product modifications to meet their particular requirements consume precious engineering resources. The more the company accommodates such requests for specials, the more the company takes on the aspects of an engineering consulting firm rather than a manufacturing company. When important customers make such requests, they may have to be accommodated. But too often technology-based companies drift into producing increasing numbers of specials when such activity is clearly not consistent with their overall strategy.

The balanced allocation of engineering resources, assuring adequate attention to the development of truly new products, is an important challenge to general managers. When the dominant view in the councils of management is production, excessive investment in product maintenance engineering will result. When the dominant view is marketing, excessive pressure for accommodating customers' requests for specials may result—to be followed soon by dissatisfaction at the slow pace of new product development. When the dominant view in management councils is development, essential product maintenance engineering may be shortchanged and very attractive opportunities for specials may be overlooked. No such myopic views can be permitted to dominate.

Balance the allocation of engineering resources among product maintenance, specials, and development.

∴Communication between production and engineering is particularly intense, and often necessarily nonroutine, in connection with introducing new products onto the production floor from the engineering laboratory. Prototype and pilot production runs can assist in the transfer, as can mutual agreement on timely freezing of the design. The more dependent the company is on process technology, rather than state-of-the-art product technology, the more thorough must be the product and process documentation. Because documentation is both expensive and difficult to control, high-technology companies typically underemphasize it. To maintain careful control of products, processes, quality, costs, and inventory investments, you must subject suggested changes in existing products to strict and well-defined procedures to be certain that the myriad potential ramifications of the change are fully evaluated. In formulating its engineering change policy, the high-technology company should consider the trade-off between introducing a new model and permitting product evolution by means of a series of incremental changes. The policy must also assure that engineering resources are not so committed to product maintenance and customer specials that new product development is shortchanged.

The Optimum Degree of Vertical Integration for a High-Technology Company

Some high-technology companies are largely self-contained and self-sufficient—that is, they are highly integrated—and others rely heavily on vendors for both components and processing. For example, Tektronix in Beaverton, Oregon, has traditionally produced in-house virtually every part in its oscilloscopes, from the important cathode-ray tube to the seemingly trivial buttons on operator panels. Many electronic instrument companies, however, pride themselves on doing no production other than final assembly; they subcontract all fabrication, processing, and subassembly work (particularly the assembly of printed-circuit boards). Some even subcontract the final assembly work.

Vertical integration implies self-sufficiency.

In the early days of small computers, many companies incorporating computing capability into their products or systems decided to backward integrate (that is, back toward raw material sources) into the design and manufacture of their own minicomputers. Involving more than a simple make or buy decision, backward integration implies technological self-sufficiency. As the minicomputer industry has matured, more of these companies have decided to rely instead on the large, specialized minicomputer manufacturers to supply computing capability on an OEM basis. In effect, they have reversed the process of backward integration.

The term *vertical integration* is commonly used to refer only to the extent to which a company's management facilities are self-sufficient. A fully integrated company starts with the raw material (for example, iron ore or sand) and produces a finished product, which it then sells directly to the end user. Virtually no high-technology companies are fully integrated.

A comprehensive view of vertical integration concerns all of the following:

1. Classic backward integration: producing parts or components that could alternatively be purchased from suppliers or subcontracted (competing with one's suppliers)

2. Forward integration in product: incorporating a company's components, equipment, or services in a larger system to satisfy a broader set of customer needs (competing with one's customers)

Remember that product and manufacturing integration implies technological integration.

3. Forward integration in marketing: assuming some of the sales, marketing, distribution, and servicing activities that could alternatively be left to the distribution channels (competing with one's distribution channels—discussed in Chapter 2)

4. Forward or backward integration in technology: assuming responsibility for progress and evolution of certain technology (both product and process) that will be competitively significant (When a company backward or forward

integrates in either manufacturing or product design, it typically is also integrating with respect to technology.)

All forms of vertical integration are important in high-technology companies. Like product-process linkages and engineering-production communication, integration must be considered in the context of the high-technology company's technical policy.

Vertical integration is centrally important to implementing the firm's technological policy, as well as its overall business strategy. Both engineering and manufacturing play important roles in that implementation.

Integration Pros and Cons

The advantages of forward and backward integration are both widely recognized and highly seductive.

Don't let the obvious advantages of integration cause your firm to overlook the subtle disadvantages.

1. Capture profits that would otherwise flow to suppliers or distribution channels
2. Improve deliveries by eliminating dependence on third parties
3. Improve quality (The key and often fallacious assumption behind this statement is obvious.)
4. Eliminate dependence on a scarce and valuable capability held by outsiders
5. Gain additional capability that you may be able to use strategically in the future to gain a competitive advantage

The disadvantages and risks, however, are very real and potentially costly.

1. Eliminate certain flexibility that is inherent in being able to deal with a variety of suppliers and marketing outlets
2. Create imbalances in capacity and capability if some portions of the business are much more integrated than others
3. Draw the company out of the niches in which it enjoys distinct competitive advantages (Corollaries are possible dilution of management attention and strain on management's capabilities and efforts.)

4. Dilute the company's overall return on investment if the activities into which the company integrates are inherently less profitable (because of competition) than the core business

5. Increase operating leverage and the company's break-even volume of operations and thus increase financial risk

6. Cause management to overlook possibilities to manage suppliers, marketing outlets, and customers in ways that would achieve some of the advantages of integration at lower total costs

Implications of Integration

The extent of integration within a company has important and mutual implications for both manufacturing and engineering. For manufacturing, the more integrated, the larger and more diverse must be the fabrication, processing, and assembly activities the company undertakes. But a commitment to backward integration—to producing in-house a certain part or subassembly that has been or might be procured from outside vendors—commits the engineering departments as well. Both design engineering and manufacturing engineering must support this decision by originating the part or subassembly, designing and documenting it in detail, and reengineering it as required to incorporate new techniques or materials and to reduce its manufacturing cost.

Manufacturing's commitment to backward integration also commits engineering.

What factors influence this decision? Leaving aside institutional or top management pride or ego, which sometimes argues for excessive integration, the decision is arrived at by answering the following six questions:

- How key is the relevant technology?
- Are you willing to commit to the technology?
- How large is the volume likely to be?
- How large are the investment requirements?
- Should operating leverage be increased or decreased?
- Are reliable and responsive subcontractors available?

Integrate into technologies and know-how that can result in competitive leverage.

Relevant Technology. An important competitive advantage for most high-technology companies derives from their detailed knowledge of and leadership in one or more specific technologies. Managers in high-technology companies should carefully analyze their present and probable future products to isolate those technologies that now provide, or are likely to provide, distinct competitive competencies for the company. In many instances, clever techniques, based upon trade secrets—often called *know-how*—may be a more important source of competitive advantage than patent-based science. The company needs to control those portions of its products or systems that embody these technologies or techniques.

For example, a process-control company may decide that the key technologies that provide it with competitive leverage in the marketplace are certain sensor technologies and the software control algorithms. This same company may also decide that the computer central processing unit (CPU), memory, and input-output devices incorporated in its process-control system are not pivotal technologies. It will look to vendors not only for supplying these major system subassemblies but also for keeping it abreast of changes and improvements that will be forthcoming in these devices. Manufacturing's role is limited to procuring these devices and incorporating them into the system at final assembly. Engineering's role is to monitor developments by others so that the company captures the full advantage of these developments and improvements.

One robot manufacturer, supplying robots to heavy industry for use in difficult environments, may elect to concentrate on the mechanical technologies involved. Another may believe that its competitive advantage is embodied in improved electronics. A third may concentrate on the challenging control problems faced in applying robots to assembly operations. Each of these companies should integrate with respect to different elements of the robotic system and therefore stress different talents on its engineering staff. The first should avoid subcontracting the mechanical portions of the systems. The second should avoid subcontracting the electronics. The third may be willing to buy "on the outside" both mechanical and electronic subsystems, concentrating its attention on software.

Security of product design and process know-how is also enhanced by in-house manufacture. The potential compromise to security argues against subcontracting: Communicating key design concepts to vendors increases the probability of proprietary information being leaked to competitors. Sometimes mere knowledge that a new generation of a high-technology product is under design within a company may be useful to competitors.

Commitment to the Technology. As a corollary to the first issue, managements of technology-based businesses should be certain that they are willing to make a long-term commitment to stay up to date in the technologies that they elect to maintain in house.

The company that elects to build, rather than buy, minicomputers for incorporation into its products must commit itself to substantial investments in computer development to ensure that the company remains current in minicomputer technology. The apparent unit cost advantage associated with in-house manufacture often evaporates when the company carefully evaluates the engineering commitment it must make, not only to improve its minicomputers' performance but also to reduce their cost.

Does the necessary and continuing investment in the technology outweigh the manufacturing cost advantage?

When Tektronix elects to produce plastic push buttons for its operator display panels, it commits engineering to remaining current not only on the availability of new plastics for buttons but also on improvements in techniques for operators to interface with display panels. Manufacturing must stay current on new plastic-molding techniques and any other processing technologies that may improve the performance or reduce the costs of the buttons. If unwilling to make these commitments, Tektronix should look to outside vendors for these small parts for its instruments.

Companies that are either at a great competitive advantage or disadvantage in terms of technological competence should consider technology *dis-integration*. The firm that competes in technology-based product-market segments and finds itself lagging in technology may decide to look outside the

If your company has a competitive advantage or disadvantage in technology, consider dis-integration.

organization to shore up its technological competencies—that is, to dis-integrate. Acquiring licenses, buying patents, entering joint ventures, and subcontracting the design of technical components or subsystems previously designed by the company reduce backward integration. Buying a standard component or subsystem from normal industry suppliers, rather than producing it oneself, has the effect of subcontracting technological development and design. The company should also consider contracting with research laboratories, including universities, when these laboratories have special competence in a technology that is (or might prove) critical to the company.

The company with extraordinary technological capabilities—capabilities that are not fully consumed by the product-market segments in which it competes—should consider licensing its technology to others, entering joint ventures, and accepting contracts for design. These and similar actions permit the company to capitalize on its technological strengths without forward integrating into manufacturing and marketing the resulting products or services.

Volume. A part or subassembly used in large volume is a better candidate for backward integration (that is, in-house manufacture) than one required in only limited quantity. The appeal of backward integration is to capture the profit that would accrue to the vendor if the part were purchased outside. However, a vendor enjoying large sales volumes, and therefore substantial accumulated experience, may be able to offer components at prices that are well below its customers' anticipated manufacturing costs.

Is volume sufficient to achieve learning curve economies? Thus, a high-technology company must anticipate enough volume in any component it elects to produce in-house to permit it to move rapidly down the experience curve, thereby realizing improved unit costs on that component. But recall that further improvements in cost come slowly once large volumes are achieved. Some cost disadvantage vis-à-vis vendors can often be tolerated in order to achieve other advantages associated with in-house manufacture.

The volume of integrated circuits most high-technology companies consume does not warrant in-house manufacture. These companies rely on so-called merchant houses: companies whose business is the design and manufacture of semiconductors and integrated circuits. There are exceptions, however. IBM, the telephone company, and Hewlett-Packard consume enormous numbers of semiconductor devices, such large quantities that in-house manufacture is clearly warranted. These companies suffer no cost disadvantages vis-à-vis the merchant houses; moreover, because they capture the manufacturing profits on the semiconductor devices, they typically enjoy cost advantages over their competitors in the computer and telecommunications businesses. These companies also have strategic reasons for being strongly committed to the many technologies involved in designing and manufacturing solid-state devices.

Investment Requirements. The simple magnitude of the investment required may preclude the alternative of backward integration, particularly because high-technology companies are frequently small and undercapitalized. Stated another way, the high-technology company may have so many attractive investment opportunities that promise high returns—particularly investments in the development of new or extended product lines—that investment in manufacturing facilities to produce in-house what it could purchase on the outside is simply not justified.

As the business matures, its risks moderate and its cost of capital declines. Investment in facilities to backward integrate then becomes more attractive. For example, the investment in a sophisticated printed-circuit board facility that promises a 30 percent after-tax return on investment may be an unattractive investment opportunity at one stage of a company's life and a very attractive one at another stage.

Don't squander limited financial resources on low-return facilities.

Some manufacturing facility investments seem never to offer attractive returns; backward integration into these activities is generally foolish. Three common examples are plastic molding, printed-circuit board assembly, and metal machining. In many metropolitan areas, independent subcontractors for machining, plastic molding, and PCB stuffing abound. Competition among them is keen; prices are relatively low; and quality and delivery are acceptable. This condition seems to occur for at least two reasons. The owner-managers of these subcontracting shops want to be in business for themselves— that is, the decision to enter business is as much a life-style decision as an economic one. Relatedly, the barriers to entry— money and know-how—are low, thus encouraging new entrants and assuring frequent overcapacity conditions in the industry. Whatever the reasons, the condition argues for utilizing such subcontractors rather than backward integrating into these processing facilities.

Operating Leverage.   Integrating vertically increases the firm's operating leverage. *Operating leverage* refers to the relationship between costs that vary with volume and those that are fixed regardless of volume. I discuss the implications of operating leverage in greater detail in Chapter 6, but it is important to recognize here that as fixed costs increase in connection with vertical integration, operating leverage increases. Increased operating leverage leads to greater risk of loss in slow economic times but higher profits in prosperous times.

The cost of acquiring a part or component from an outside vendor is a variable cost; it is incurred only when the part is purchased, and the total cost varies directly (ignoring quantity discounts) with the number of parts purchased. In-house manufacture of the same part inevitably exposes the company to the fixed costs associated with the required facilities and production management; that is, some costs incurred are independent of the volume of the part produced.

When you seek the advantage of increased operating leverage, backward integration is an attractive way to achieve it. When you wish to avoid the risks associated with increased operating leverage, subcontract.

Balance risks and keep costs variable.

A key general management task is to balance risks. Most high-technology companies, particularly small ones, are already exposed to considerable risk. The risk inherent in rapidly changing technology is a primary one, but considerable financing risk (discussed in Chapter 7) is also typical. Where technological and financial risks are high, operating risk—that is, operating leverage—should be minimized. As a result, most smaller technical firms should have an established policy of subcontracting—that is, relying on outside vendors—whenever possible. The guiding rule is to keep costs variable.

Subcontractor Availability. This discussion implies that management has a choice: that subcontractors or vendors are available. The luxury of having a choice may be a function of where the company is located.

One of the compelling motivations for Tektronix to backward integrate so thoroughly in its early years was the absence of viable subcontractors in close proximity to the company's headquarters in Beaverton, Oregon. The infrastructure, to borrow a word from the economists, of the region in which the company is located must influence the company's backward integration policy. A key advantage accruing to the start-up company in the San Francisco Bay Area, Los Angeles, or the Boston area is the existence of an elaborate infrastructure serving high-technology companies. Besides a broad array of professional services, subcontractors of a wide variety exist: metal fabricators, design job shops, software developers and consultants, plastic molders, testing laboratories, PCB assemblers, and temporary help agencies. (The use of temporary help, from engineers to machinists, assemblers, and clerks, can be thought of as a kind of subcontracting.)

Does your region's infrastructure allow you to avoid backward integration?

Sometimes the mere existence of subcontracting capability is not sufficent. A company's demand for fast delivery, exceptional quality, or very close interaction between engineering and manufacturing may preclude the use of other than company-owned facilities. Many companies that elect to subcontract should also maintain a prototype capability (for example, machine shop) in order to assure very fast turnaround on newly designed parts and to facilitate engineering communication with a minimum of formal documentation.

Forward and Backward Integration

I have been speaking largely of backward integration—that is, producing components or subassemblies of existing products. Forward integration is incorporating the company's primary products or devices into more comprehensive products or systems. It involves competing with one's customers, which has some negative implications. Customers are almost sure to be displeased. If their displeasure is sufficient, they may switch to suppliers who avoid this head-to-head competition. Nevertheless, when the supplied component becomes a major—or *the* major—determinant of the competitive positioning (both cost and performance) of the downstream product or systems, the advantages of forward integration can be overwhelming.

Consider forward integration when your component is key to the downstream product or system.

The entry of Texas Instruments and other semiconductor manufacturers into the production of low-cost, handheld calculators and of digital watches is a case in point. Competing companies that relied on merchant houses for the key components—electronic chips—for these low-priced consumer items were at a hopeless cost disadvantage vis-à-vis the large, integrated manufacturers, such as TI. The very aggressive pricing by Texas Instruments—pricing in anticipation of (and to achieve) very large volumes and thus scale economies—simply drove out the nonintegrated suppliers in these industries.

I discussed in Chapter 2 the choice many high-technology companies face: selling direct (deploying their own sales force) or selling through reps (using commissioned agents). Selling direct involves a form of forward integration. The considerations discussed previously regarding backward integration apply equally to this marketing decision:

1. How key is sophisticated technical selling in the overall business strategy? Selling, including application engineering, may be an important basis of differentiating products and services from those of competitors.
2. Is the company willing to commit the management effort to hiring, training, supervising, motivating, rewarding, and supporting a direct sales force?
3. Is volume sufficient to justify the fixed expenses inherent in direct selling?

4. Can the company afford the investment in a direct sales force?

5. Can the company tolerate the increased operating leverage inherent in direct selling?

A particular advantage of forward (marketing) integration in high-technology companies is improved access to market and competitive information. When products and markets are changing rapidly, the company must organize to obtain and systematically process information regarding those changes and competitors' responses. Direct selling aids this process.

View direct selling as a form of forward integration.

Partial Integration

Partial or tapered integration, either forward or backward, a half step toward vertical integration, involves the company in producing some of its requirements and purchasing the balance on the open market. The company's requirements must be sufficiently large to sustain an efficient internal operation. Partial integration may mitigate the disadvantages of high fixed costs and capacity imbalances that frequently accompany full integration. It may also place the company in a strong bargaining position with its suppliers and provide it with some benefits of information, rapid turnaround, and internal competencies that attend full integration.

Vendor Relationships

The less vertically integrated the company, the more it depends upon its suppliers and the more it should devote attention to articulating and implementing a coherent vendor relations policy. In Chapters 2 and 3, I discussed OEM relationships from the viewpoint of the seller, or supplier. Now the shoe is on the other foot. The viewpoint is that of the customer, or the OEM.

Negotiations between customer and vendor should encompass price, terms of sale, delivery, quality, and technical assistance (both pre- and postdelivery). Inexperienced buyers tend to focus solely on price and delivery. Sophisti-

Avoid adversarial relationships with your suppliers.

cated purchasing executives seek a broad set of negotiating parameters, attempting finally to arrive at an agreement that provides the optimum mix of advantages to both parties. An adversarial relationship between the company and its suppliers should be avoided, or at least minimized, and the mutual dependency of the two companies should be acknowledged. Again, we can take a lesson from the Japanese. The close working relationships between the large industrial companies in Japan and their legions of subcontractors on whom they depend have greatly benefited that economy. (Admittedly, the advantages seem to accrue particularly to the large Japanese companies and less to the subcontractors.)

OEM contracts in high-technology industries are particularly important. In Chapter 3, I noted that the balance of the advantage in blanket order contracts seems to accrue to the buyer. The primary advantages to the buyer are as follows:

1. Price remains fixed for the contract period.
2. Specifications and supply are assured for the contract period.
3. Additional assistance in the form of access to technology or preferred deliveries can be expected from the supplier.
4. Purchasing expenses are reduced; blanket order releases are routine.
5. Shorter lead times permit a reduction in inventory investment.

The primary penalty to the buyers is a certain loss of flexibility: they forgo the opportunity to obtain products with improved operating specifications or lower prices from alternative vendors during the contract period. For a product undergoing rapid technological change, this loss of flexibility may be significant.

Engineering must share with manufacturing the responsibility for vendor relationships.

Typically, the primary responsibility for vendor relationships—that is, purchasing or procurement—rests with manufacturing. In some companies, however, where purchasing and vendor relationships are both extensive and key to the success of the business (for example, airframe manufacturers), procurement may report to the general manager on a basis parallel with production. In any case, close cooperation between purchasing and engineering is essential in high-technology companies to be certain that:

1. Design engineers reap the benefit of technical information and interchange with personnel from vendors
2. Engineering's insights as to product specifications and vendor capabilities are incorporated in the vendor selection decision
3. Engineering participates in any decision to modify product or processing specifications when such modifications can result in lower prices, more rapid delivery, or some other advantage to production
4. Negotiations with vendors as to price, terms, and so forth occur before engineering has committed to the use of the component or subsystem

Effective purchasing—an activity that extends well beyond simple price negotiations—including close working relationships with vendors, offer profit improvement opportunities that many U.S. high-technology companies are currently overlooking.

Effective purchasing negotiations encompass more than price, and delivery terms.

∴ The firm's technical policy also serves as the basis for deciding just how vertically integrated the company should be. Vertical integration has important implications for engineering and manufacturing. Vertical integration decisions turn on (1) the importance of the product or process technology to the firm's business strategy, (2) the firm's willingness and ability to commit to the technology, (3) the volume to be produced, (4) the investment requirements to effect the integration, (5) the role of operating leverage in the firm's financial strategy, and (6) the availability of subcontractors. An analogous evaluation process applies to forward integration—into downstream products or distribution channels—in an analogous manner. Technical companies should consider both partial integration and conscious dis-integration. The less integrated the firm, the more it needs to pay close attention to its vendor relationships through competent and sophisticated purchasing.

Highlights

- Both process and product technology offer possible competitive leverage to the high-technology firm.

- An explicit technical policy is the basis for engineering and manufacturing strategies.

- Technical policy links product and process technologies, determines whether the firm will design to performance or cost, and defines the roles of design engineering and manufacturing engineering.

- When new products move from the engineering laboratory to the production floor, communication between production and engineering is particularly intense and nonroutine.

- Prototype and pilot production runs, as well as timely freezing of the new design, can help smooth the transfer process.

- Strict engineering change procedures are necessary to control products, processes, quality, costs, and inventory.

- Engineering strategy should focus on the choice between letting a product evolve through a series of incremental changes and introducing a totally new product.

- Engineering resources must be invested in a balanced manner among new product development, product line maintenance, and customer specials.

- Vertical integration commits engineering as well as manufacturing to greater self-sufficiency.

- In choosing the extent of integration, consider the importance of the relevant technology, your willingness to commit to the technology, the volume you will produce, the investment requirements, the role of operating leverage, and the availability of subcontractors.

- High-technology companies should also consider forward integration, partial integration, and dis-integration.

- The less integrated a firm, the more attention it must pay to its relationships with its vendors.

Notes

1. For a discussion of technical policy, see Modesto A. Maidique, "Corporate Strategy and Technological Policy," HBS Case Services, Harvard Business School, Boston, Massachusetts.

2. J. Fred Bucy, president and chief operating officer, Texas Instruments, Inc., November 17, 1976, New York University Key Issues Lecture Series, lecture entitled "Marketing in a Goal-Oriented Organization."

3. For a discussion of PERT techniques, see Roger W. Schmenner, *Production Operations Management* (Chicago; Science Research Associates, 1981), Appendix to Chap. 11, p. 279.

4. For a discussion of these and other issues relating to vertical integration, see Michael E. Porter, *Competitive Strategy* (New York: Free Press, 1980), Chap. 14.

Chapter Five

Assuring Quality in High-Technology Products

This chapter covers the following topics:

- The Truth About Quality: An American Problem
- Contrasting Views: Traditional and Enlightened
- Implementing the Enlightened Assumptions: A Crucial Step
- Quality Circles: An Enlightened Organizational Device

Quality assurance, long thought to be the sole province of production, is now recognized as a companywide responsibility. In high-technology companies, product quality must be specified by marketing, designed in by engineering, built in by production, and maintained by field service. Improved quality demands effective interaction and communication among the three functions on which Chapters 2, 3, and 4 focus: marketing, engineering, and production.

Currently, quality assurance is looming as a major—and in some cases, *the* major—issue for high-technology companies, as the worldwide supremacy of America's high-technology industries is severely challenged, particularly by the Japanese. That challenge turns, to a major degree, on quality differences.

"Made in Japan" was for many years used derogatorily in the United States. It implied low quality and poor workmanship. However, the Japanese people, including the government and the major industrial groups, made changing that image a national priority in the early 1950s. They have succeeded to a startling degree: Japanese high-technology products now have a worldwide reputation for outstanding quality. Interestingly, the Japanese learned the techniques for improving quality from the Americans. They have adapted and implemented those techniques with careful planning and follow-through in a culture that seems particularly receptive, a culture that emphasizes teamwork and pride. American high-technology companies can and must adopt both the commitment to and the techniques for improved quality if they are to continue to compete effectively in world markets.

Achieving higher quality requires drastic alterations in management's traditional view of the causes of poor quality, the costs and benefits of higher quality, and the responsibilities for implementing tighter quality standards. In a high-technology company, quality assurance can be a source of strategic advantage, as can marketing or engineering. In exploring these issues in this chapter, I focus on some popular misconceptions about quality; suggested changes in managements' views regarding quality; and some important techniques of quality assurance, including quality circles, an organizational device that is sweeping high-technology companies.

◆ *The Truth About Quality: An American Problem*

Major stumbling blocks for U.S. high-technology companies in their pursuit of higher quality are managers' misunderstandings of the importance of quality, the trade-offs between quality and cost, and the sources of and responsibilities for present quality conditions.

Wizardry Versus Quality

Managers in companies employing state-of-the-art technology tend to relegate quality issues to secondary importance. Indeed, in the early years of a new technology, innovation and improvement in performance specifications may carry the day. Market leadership and financial returns accrue to the company or companies that push forward the frontiers of the technology. Product malfunctions are both expected and tolerated because of the newness of the technology. In any case, customers have limited information with which to differentiate among suppliers on the basis of quality. New customer segments are tapped as improved performance is brought to the market. During these years, price, cost, and quality considerations do not dominate the competitive arena; technology does. The technological wizards are the stars.

As the new technology matures and more competitors enter the market, differentiation among products on the dimension of technical specifications lessens. As customers gain familiarity with the product or system, they are increasingly able to judge the product's total life cycle cost—the cost to use as well as the cost to acquire—and distinguish among competing suppliers on this basis. Price, and thus cost, takes on increased significance. Now quality, or reliability, can no longer take a back seat to the glitter of new technology.

The importance of quality increases as new technology matures.

Costs of Poor Quality

Managers tend to underestimate by a wide margin the true costs of quality. There are some obvious costs of poor quality: scrap, rework, warranty repair. The less obvious costs may be considerably more important:

1. Work flow disruption as material or components are rejected and returned to vendors
2. The cost of rescheduling, renegotiating with suppliers, expediting replacement parts, and troubleshooting on the manufacturing floor
3. Excessive investment in inventories to provide safety stock in the event of quality problems
4. Excessive field service costs, whether borne by the customer or the supplier
5. Most important, customer ill will that can lead to lost business in the future

Don't underestimate the costs of poor quality.

The costs are high: Some experts have estimated that poor quality costs the average American manufacturer 16 to 20 percent of sales. The profit leverage associated with improved quality is, therefore, extraordinary. If quality could be improved so that this percentage is reduced to 2 to 3 percent of sales— a realistic target, according to these same experts—about 15 percent of sales could be added to operating profit. Thus, you should view improved quality not as an added cost but rather as an opportunity for substantial profit improvement.

The Cost-Quality Relationship

Historically, managers have assumed that increased quality leads to higher costs, the trade-off between cost and quality is inevitable, and very high quality can be obtained only at very high cost. This traditional view is illustrated in Figure 5-1. When the only route to improved quality is viewed as requiring more extensive (and expensive) inspection and additional rework, this trade-off is accurate.

Improved quality results in lower, not higher, costs.

When changes in worker attitude, product redesign, process change, and other problem-fixing steps are admitted as important routes to improved quality—the steps we explore in this chapter—the family of trade-off curves shown in Figure 5-2 better describes the possibilities. Improved work force management can shift the curve, as can, for example, new process design. For a specific process design, the trade-off between cost and quality continues—more inspection and more rework can result in higher quality—but still further improvements in the process may permit a further downward shift of the curve. A series of these changes over time offers the opportunity to improve the quality-cost ratio. (Note the arrow running through this family of curves in Figure 5-2.)

Figure 5-1. Cost versus quality: traditional view

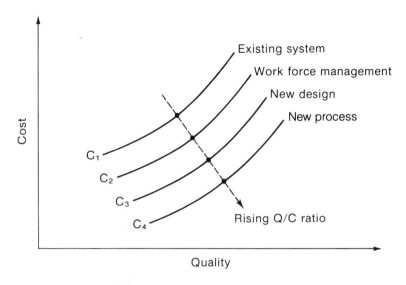

Figure 5-2. Cost versus quality: expanded view

Companywide Responsibility

Quality has traditionally been viewed as manufacturing's responsibility. The quality control group reported to the firm's senior manufacturing executive, and this group was held accountable for the level of quality of products leaving the factory. Senior management's view of quality emphasized control—quality control—or inspection. Quality control was a checking-up function and inspectors were police officers.

In fact, quality cannot be inspected into a product. It must *You can't inspect* be built in. Before high quality can be built into the product, *quality into a* it must be designed in. If product design is inherently unre- *product.* liable or calls for tolerances or techniques that are difficult to achieve in manufacturing, products of inadequate quality will probably flow from manufacturing despite the best efforts of the production force. Thus, improved quality is at least the shared responsibility of engineering and manufacturing.

But the responsibility for quality extends to other functions of the business as well. It starts with top management. The company attitude toward quality is a reflection of top management's attitude. If top management subordinates quality objectives to concerns for stepped-up deliveries or for meeting short-term earnings-per-share objectives, managers and workers within each of the functional departments will respond to this ranking of priorities. If the rewards go to those who meet shipment goals, regardless of quality, rather than to those who achieve quality targets, quality standards will be relaxed to meet shipment objectives.

Quality is everyone's responsibility.

The responsibility for quality extends also to marketing, field service, and the staff functions of maintenance, accounting, industrial relations, and so forth. Marketing plays an important role in discerning and communicating to the balance of the organization the customers' true requirements, including in-use reliability. Marketing must be involved in the

Table 5-1. Contrasting quality assumptions

Traditional	Enlightened
Quality control (inspection)	Quality assurance
Technicians (inspectors)	Managers
Inspect for failures	Prevent failures
Defects lead to blame, excuses, justifications	Defects lead to problem solving
React to problems	Anticipate problems
Defects should be hidden	Defects should be highlighted
Defects are predominantly hourly worker caused	Defects are predominantly management caused
Manufacturing oriented (product)	Organizationally oriented (process)
Operational focus only	Combined product design, process design, and operational foci
Functional, formal	Matrix, informal
Quality *versus* manufacturing	Quality *and* manufacturing
Cost/delivery *or* quality	Cost/delivery *and* quality
Quality department has quality problems	Organization (R&D, mfg., marketing, purchasing) has quality problems
General managers not evaluated on quality	Quality performance part of general manager's review
AQL (acceptable quality level)	ZD (zero defects)

internal negotiations as to acceptable quality standards and any deviations from those standards. Quality is also influenced by the maintenance standards that the company applies to its equipment and processes. Quality in customer relations extends even to the accuracy, timeliness, and courtesy with which the accounting department serves the customer.

The standards of quality to which top management commits the company become reflected in the performance standards of all its functions. The experience of companies that have placed high priority on quality has been that top managers' zealous or missionary devotion to quality gets transmitted throughout the firm. In turn, this quality thrust results in improved performance in all corners of the organization.[1]

Top management commitment is critical

∴ The traditional view of quality assurance—that it is synonymous with inspection—are invalid. High quality need not necessarily mean high cost. Improved quality and lower costs can be simultaneously pursued through better product design, process design, and work force management. Responsibility for good quality, which starts with top management, belongs to everyone in the firm.

 # *Contrasting Views: Traditional and Enlightened*

A first step toward improving quality is to change people's misconceptions about it. Quality must no longer be thought of as inspection. A more enlightened view of quality emphasizes quality improvement—quality assurance—a management function. Table 5-1 elaborates the contrasting viewpoints—traditional versus what I call enlightened. Most Japanese firms and an increasing number of U.S. high-technology firms have adopted a more enlightened view of quality.

High-technology firms must adopt the enlightened view of quality.

Unfortunately, the view that I label traditional continues to prevail in many mature industries in this country, where management attitudes tend to be autocratic, and union-management relationships are antagonistic and adversarial. They cannot be permitted to prevail in high-technology companies.

Try to prevent failures, not just fix them after they occur.

The objective of a well-managed quality assurance function is to prevent failures—to discover the causes of the quality problems and eliminate them. It is not enough simply to find the bad product and reject it; such limited activity leads to blame, excuses, justifications, and an adversarial relationship between inspectors and line manufacturing personnel. When the more enlightened assumption—that quality assurance seeks to prevent failures rather than simply to find and report them—is made, the discovery of defects leads to problem solving rather than finger pointing. Inspection helps manufacturing improve operations rather than simply thwarting its efforts to ship the product. Quality assurance personnel become proactive intermediaries, working with all functions of the company to improve quality. Defects are highlighted so that problems can be solved rather than being swept under the rug.

This enlightened view of quality also recognizes that the primary responsibility for poor quality—and therefore for quality improvement—rests with managers and not with hourly workers. Quality attention has traditionally been focused on manufacturing and on the final product. Increasingly, attention is being broadened to all parts of the organization, as management recognizes that both product design and process design play particularly key roles in quality improvement. Thus, quality management is not limited to the functional and formal activities of inspection, but instead involves a multi-functional and yet informal approach to problem solving.

Make zero defects your goal.

Quality standards should not be static. Rejection of a certain percentage of output need not be accepted as the norm. Enlightenment includes realizing that an absence of errors is the real target. Instead of an acceptable quality level—an AQL of 1 or 2 percent—representing good performance for the organization, the objective becomes zero defects—"do it right every time." An interesting manifestation of this shift in viewpoint is the Japanese shift in quality definition from AQL—

percentage defect—to PPM—defective parts per million pieces produced. (A defect rate of one hundred parts per million is equivalent to 0.0001 percent defect.) The change in definitional terms is subtle but powerful.

Some of the old adversarial relationships and assumed trade-offs must change if a company is to move from the traditional to the enlightened set of assumptions. Instead of the quality control people policing manufacturing's output—a task definition that is bound to lead to an attitude of quality *versus* manufacturing—the quality assurance group assists manufacturing in defining problems and developing solutions.

Quality assurance assists, and does not fight, manufacturing.

∴ The expanded view of the cost-quality trade-off shown in Figure 5-2 requires the broader view of quality that is implied by this enlightened set of assumptions. No longer are quality problems laid solely at the feet of the quality organization. Quality improvements are seen as the mutual responsibility of all parts of the organization, and particularly development, line manufacturing, purchasing, and marketing. The general manager's performance must be judged in part on quality. The enlightened assumptions are more likely to prevail if quality commands an important position on the list of criteria by which the general manager's performance is evaluated. When the general manager believes that meeting quality standards ranks on a par with meeting delivery and near-term profit goals, the balance of the organization will similarly place high priority on improved quality.

➤ *Implementing the Enlightened Assumptions: A Crucial Step*

If quality is to be improved, more is required than simply willing a new set of assumptions on the organization. Two American experts on quality who are widely quoted in Japan,

W. Edward Deming and J. M. Juran, have some simple but powerful suggestions. In the early 1950s, the Japanese government invited Edward Deming to come to Japan to assist the national effort to improve quality. The prestigious annual Japanese award for quality excellence is now called the Deming Award.

The contrasting assumptions shown on Table 5-1 encapsulate much of Deming's and Juran's thinking and teaching. Only recently is their advice being heeded at home as it has been abroad for thirty years. Several action steps follow from the enlightened set of assumptions:

1. Top managers must accept responsibility for quality.
2. Sampling is necessary but not sufficient.
3. Testing should be done early and often.
4. Quality assurance should be made a staff role.
5. Quality problems should be exposed and highlighted.

Top Management's Responsibility

Don't pressure workers to correct systemic problems.

Top managers must accept the primary responsibility for quality improvement and must demonstrate their commitment by tangible and apparent actions. Deming argues that 85 percent of quality problems are management's responsibility. Placing increased pressure on workers for improvements in quality can, under these circumstances, prove counterproductive. When product or process design, poor maintenance, or inadequate tooling, for example, is the source of the defects in production, no amount of pressure—whether it takes the form of incentives or disciplinary threats—can improve quality. The workers will only encounter frustration as they attempt to meet management's unreasonable expectations.

Thus, management must be the starting point for any conscientious program for quality improvement. Managers need to fix the systemic problems—problems that are beyond the control of the production workers—before the workers can be effective in joining the battle for improved quality. Stepping up equipment maintenance, replacing aged and worn equipment, improving lighting and plant cleanliness, and committing engineering time to the redesign of particularly troublesome components are often useful steps.

Top managers must be willing to listen and learn from operating personnel just what systemic problems need correcting. As quality improves, top managers must highlight these improvements, just as they highlight improvements in order bookings, shipments, and earnings per share. Top managers must accept quality as one of their own performance criteria.

Evaluate top managers on their quality achievements.

Statistical Sampling

Deming argues that the traditional preoccupation with statistical sampling is inappropriate when you are seeking the root cause of quality problems. Only 100 percent testing can reveal the various causes of the defects and also their relative frequency of occurrence.

Thus, sampling is appropriate for monitoring a process to see that it remains in control; but sampling implies that a certain percentage defective is acceptable. When the objective is shifted from acceptable quality level to zero defects, the cause of any defect—that is, any unacceptable product—must be discovered. Sampling is a waste of time and money in the hunt for appropriate corrective steps to prevent failures. One hundred percent testing should be encouraged, together with careful documentation of results, to discover causes of problems so that the problems can be corrected.

Encourage 100 percent testing to find causes of poor quality.

Testing

Testing frequency and location are critical for three reasons. First, the earlier in the process that the problem is uncovered, the lower the cost of the poor quality. If defects are discovered in raw materials or components, the cost of rework or scrap may be only pennies. If this same defect is not discovered until customers are using the final product, the cost of effecting repair or replacement may run into hundreds of dollars.

Test early and often.

Second, if testing is accomplished immediately following the performance of a particular manufacturing step, feedback is rapid and direct and therefore particularly useful. The operator (or assembler), rather than an inspector, should perform the test. This operator is in the best position to interpret the significance of the test results and to take any corrective measures that may be indicated. The operator should be encouraged to call for help from others when and as needed. Management's obligation is to be certain that assistance is available when called for so that the operator is not frustrated in his or her attempts to improve quality.

Third, if testing is early and often—immediately following each step in the process—necessary corrections can be made before much more bad product is manufactured.

Role of Quality Assurance

Quality assurance assists rather than inspects.

This call for frequent and timely inspection requires that the responsibility for inspection be with line manufacturing personnel, not relegated to a set of quality control inspectors whose job is to check up on production personnel. Quality assurance is a staff role to (1) assist the line organization in designing appropriate test procedures, (2) establish quality standards, and (3) assist and coordinate the various functions of the organization in attempts to correct the problems so as to eliminate defects.

In stressing the importance of top management's commitment, Juran offers an interesting suggestion to middle managers—particularly quality assurance managers—in gaining this commitment. He suggests that the language of top management is finance, or money, but the language of operations is "things" (quantity of output, number of labor hours, and number of rejects, for example). The challenge to the quality assurance staff is to bridge the gap between these languages, to translate the cost of quality and the opportunity for improved quality into dollars and cents terms. Such reinterpretation of quality is an important step in gaining the involvement of all the functions of the company in the business of improving quality.

Highlighting Quality Problems

Many inherent quality problems are hidden from view both by individuals who fear punishment and by certain widely practiced procedures. The first cause, fear, can be overcome only when all members of the organization believe that management has adopted the enlightened assumptions. When management accepts the primary responsibility for quality, operating personnel will no longer be reluctant to bring quality deficiencies to light.

Expose and highlight quality problems.

Other steps should also be taken to expose the sources of poor quality. Gradually reduce safety stocks to the point where they no longer compensate fully for changes in yield or abnormal numbers of rejects. If quality fluctuations and problems are leading to the assignment of additional personnel (for example, on assembly lines or within the field service function), consciously and steadily reduce staffing levels. The savings in inventory investment and labor costs are both obvious and desirable, but the fundamental objective is to expose and highlight quality problems so that they can be fixed. As long as they remain hidden from view, the root causes of the problems will persist. Management should be alert to other operating practices that have become built in as an offset to inadequate or erratic quality. The systematic (generally gradual) elimination of these practices can pay substantial dividends.

As enlightened assumptions become fully accepted, the discovery of a quality defect is viewed as a cherished event, for it represents an opportunity to make changes that will, in virtually all cases, lead both to lower costs and to higher final quality.

∴ Enlightened views of quality assurance recognize that the responsibility for quality improvement starts with top managers and involves all functions. The root causes of poor quality are more likely to be systemic (process related) than worker related. The objective should be to eliminate defects, not simply to reduce them to a tolerable number. The quality assurance function should assist the line production function. The purpose of inspection, which should occur early and often, is to find causes, not just symptoms, of poor quality.

❧ *Quality Circles: An Enlightened Organizational Device*

Quality circles are an organizational device that both relies on and supports the enlightened set of assumptions. They are a Japanese invention that is rapidly gaining popularity in this country. The concept is reasonably simple, but its implementation is substantially more difficult.

Quality circles are as useful for increasing productivity as for improving quality.

A quality circle consists of a group of employees (typically seven to ten) from a particular department or work area who gather on a periodic and regular basis (typically weekly, for about an hour, on company time) to consider problems that need to be solved or improvements that might be instituted. Cross-department quality circles have also proved effective, but most companies concentrate initially on circles whose members are drawn entirely from a single department. In spite of the name, the tangible results of a quality circle program are as much increased productivity as improved quality.

Many have criticized quality circles as being nothing more than a management fad of the 1980s. It is true that many high-technology companies have rushed into quality circle programs without proper planning, expecting them to be the magic road to improved quality. Disappointments have, of course, resulted.

The QC circle concept is built on sound principles of organizational behavior, human motivation, and effective management. But implementing quality circles requires thorough understanding of their functioning, careful preplanning, active support of top management, and a good measure of patience.

Management's Role

Management must provide the framework and support for the work of the quality circles and may encourage their formation, but management cannot and must not mandate the creation of particular circles. A key precept of QC circles is that membership in a circle is strictly voluntary.

Management must support, but cannot demand, quality circles.

Among the support that needs to be given the circles is participation by a facilitator—a person trained in the techniques of quality circles, including the techniques of problem identification and brainstormng for alternative solutions. Considerable training must also be provided to the participants to help them improve their ability both to identify and to solve problems. This training should be ongoing as the circle increases its effectiveness.

Operation

The chair of the meeting may be one of the members of the circle, and this responsibility may rotate. Instead, the work area supervisor may serve as chair, the typical pattern in Japan. The process, or sequence of steps, that the quality circle follows is as follows:

1. Problem identification—developing a list of problems or opportunities for improvements within the work area
2. Problem selection—ranking the identified problems by both how important and how tractable they are (Particularly in its early days, encourage the circle to select problems that have a high probability of leading to successful solutions. Early successes have important reinforcing effects.)
3. Problem analysis—uncovering the true dimensions of the problem and alternative solutions
4. Recommendation to management—typically, quality circles make formal presentations to management, but the final decision regarding implementation remains with

management (Experience suggests that managers accept about 80 percent of the recommendations).

Make available to quality circles data for analysis.

Circles often do not have available the data necessary to analyze and evaluate problems and opportunities or the specialized skills to formulate all aspects of the solutions. A properly designed quality circle program must assure that these data and specialized resources are made available to the circles at their request. Again, active support from top management is essential.

Dangers

Developing an effective quality circle program is not simple. Considerable, long-term commitment by top management is required. A primary danger is that managers will lose patience. They cannot treat circles as a fad, a quick fix to quality problems. Workers must be trained; the support structure must be built; and facilitators must be hired and trained. Quality circles involve hard work, with an emphasis on many small details and a willingness to persist and persevere.

Long-term commitment and patience are essential.

Second, management's priorities cannot switch from improved quality of output to increased output of any quality without undermining the quality circle's program. Improved quality cannot be the objective this month, this quarter, or this year. Management must be willing to have the company evaluated on tangible measures of quality—by its customers, its shareholders, its suppliers, and its employees—just as it is evaluated on such tangible financial measures as growth rate and earnings per share. Customer communications, annual reports, and employee newsletters must demonstrate that the company has assigned quality a high priority.

A third danger is that middle managers, although capable of contributing substantially to the work of the quality circles, may instead offer resistance. This resistance is particularly severe when the quality circles uncover middle management mistakes or shortcomings.

A fourth, and perhaps the greatest, danger is that management will drag its feet in correcting the systemic causes of poor quality (for example, poor product or process design or inadequate equipment or procedures). Quality circles can point up systemic problems that they cannot correct themselves without some action by other departments or by senior managers. If management refuses to acknowledge the broader, more fundamental problems—or is slow in correcting them—circle members become frustrated and disillusioned about management's true commitment to quality.

Management should focus on identifying and correcting the most blatant of the systemic problems *before* instituting a quality circle program. If it fails to do so, the circles' early deliberations will be preoccupied by these problems, to the exclusion of problems that can prove more tractable by the group. Japanese parent firms that have acquired and now operate U.S. facilities have been slow in instituting quality circle programs in their U.S. operations. They feel that management must take certain corrective actions before quality circles can be productive.

Correct blatant systemic problems before instituting quality circles.

A final danger is that management will become preoccupied with measuring the payoff from a quality circle program. Financial evaluation is difficult; the Japanese do not even try. Nevertheless, firms that have successfully implemented quality circles are typically confident that the payoff has been handsome.

Assumptions

The concept of QC circles is deceptively simple. The assumptions upon which it is built are obvious and consistent with those I have labeled enlightened.

Any gathering of interested employees to focus on work method improvements is likely to be productive. A key assumption behind the quality circle concept is that workers are capable of, and interested in, making significant improvements in their productivity and quality of output. Workers are uniquely positioned to identify both problems and opportunities because they are intimately familiar with the work involved and because workers sharing ideas can produce creative solutions.

Workers are in the best position to identify and correct both productivity and quality problems.

The assumptions underlying quality circles are the opposite of the Taylorism assumptions that underpinned the early work in job simplification and assembly line methods. Taylorism calls for the separation of the planning and the execution of work. Staff specialists, typically industrial engineers, plan. It is assumed that lack of training and education on the part of the workers substantially limits their abilities to contribute to work improvements. If these assumptions were ever valid, they are less so in an age when workers are better educated and may bring to their positions considerable previous experience.

The obvious success of quality circles in Japan and in many companies in this country has proven the validity of the revised set of assumptions regarding workers' ability to contribute. Quality circles are now the norm in large companies in Japan; a large majority of the hourly workers participate in them, and the circles are spreading slowly to white-collar workers as well.

The name quality circle (or QC circle), although handy, is somewhat restrictive. Productivity teams or similar labels may be more appropriate and are in fact used by some companies. The point is that the benefits are by no means limited to quality; they extend also to the interwoven benefits of improved productivity, increased job satisfaction, the development of mutual respect, enhanced employee development, and quality of work life.

∴ Quality circles can be effective over the long run in engaging workers in the search for improvements in products, processes, and procedures that result in higher productivity and quality. They are based on the enlightened assumptions regarding quality assurance, but they are not a short-term fix. Implementing a quality circle program requires the commitment and patience of top management and a willingness to correct systemic problems.

Highlights

- A high-quality product is the result of good product design, process design, and work force management, not of inspection.

- High quality does not equate with high cost, but poor quality is very expensive.

- Quality is the responsibility of all parts of the organization—particularly development, line manufacturing, purchasing, and marketing—but responsibility starts with top management.

- Causes of poor quality are typically more process related than worker related.

- The objective of quality assurance is to eliminate the causes of defects rather than to reduce their number to a tolerable level.

- Quality assurance is a staff function, assisting the line organization in testing early and often to find causes, not symptoms, of poor quality.

- Quality circles, which are based on the enlightened assumptions regarding quality assurance, are effective in engaging workers in the pursuit of improved products, processes, and procedures to attain higher quality and productivity.

- Quality circles can be effective only if top managers support them and are patient for results.

Notes

1. The multifunctional nature of quality is captured in the phrase "total quality control," a phrase made popular in the 1960s by A. V. Fergenbaum in *Total Quality Control* (New York: McGraw-Hill, 1961). The phrase is now widely used among quality-conscious managers and quality assurance personnel in Japan.

Chapter Six

Understanding Financial Return and Its Effect on Company Growth

This chapter covers the following topics:

- The Pressure for Growth: Markets and Staff
- Return on Equity: Measuring Your Company's Ability to Finance Growth
- Determining the Amount of Investment Your Firm Requires to Grow
- Three Avenues to Increased ROE
- Analyzing Capital Investments in High-Technology Companies

We turn now from considering the boundaries among engineering, marketing, and production to focus on money, or finance, in this and the following chapter. This chapter builds directly on the previous discussions of the interfaces among the key functional areas of the high-technology firm. Many of the topics of previous chapters reappear as I consider their impact on the economics of the firm and on the source and allocation of financial wherewithal.

Rapid growth characterizes most high-technology companies, and it must be financed. A company's ability to finance growth internally—that is, without turning to outside sources for new capital—is a function of (1) its profitability in relationship to both sales and assets, (2) its asset intensity (that is, its investment in working capital and fixed assets in relation to sales), and (3) its debt leverage (the extent to which it relies on borrowed funds). The interrelationship between company growth rate and these three factors is the subject of this chapter.

In exploring this interrelationship, I consider the pressures for growth that exist in high-technology companies; the importance of profits—that is, return on equity (additions to retained earnings)—as the key source of financing that growth; the primary avenues for improving return on equity; the paramount importance of asset management—particularly current assets—in high-technology companies, where accounts receivable and inventories are typically the dominant assets; and caveats regarding the application of widely accepted techniques to the analysis of (necessarily) risky capital investments in high-technology companies.

⬥ *The Pressure for Growth: Markets and Staff*

Rapid market growth demands rapid company growth.

As markets are developed and the customer base for new technological products or services expands, the high-technology firm typically faces a classic dilemma: grow with the market or fall behind and lose competitive leverage. If the new technology is revolutionary, market growth—and therefore pressure for company growth—may be great.

Not all firms can or must grow as fast as the broad markets they serve. They may, for example, choose to redefine their businesses into successively smaller market niches as the total market explodes. A dominant position in a key market niche often affords substantial competitive protection. Nevertheless, companies with high market shares tend to be more successful over the long term than do those with small shares. Maintaining or increasing market share implies that a company must grow at or above the rate at which its market is growing.

Rapid growth creates opportunities for a talented staff.

A talented staff also pressures the company toward high growth. High-technology companies that grow rapidly have greater success in attracting first-rank engineers, managers, and other employees than their slower growing cousins. High-growth companies provide the challenges and opportunities for advancement that talented individuals seek. Should growth slow, these individuals may be enticed away by the greener pastures of other high-growth companies.

Thus, many high-technology companies face the need to grow at rates substantially in excess of 3 to 5 percent per year (real growth, adjusted for inflation), a rate that characterizes mature industries, or even well above a 10 to 15 percent (real growth) rate that represents outstanding performance for the best companies within those mature industries.

Financing a high rate of growth is a particular challenge for high-technology companies. Responding to that challenge is by no means the sole responsibility of the finance function. The wherewithal to finance growth comes from one or a combination of the following three sources: retained earnings, additional credit (borrowings), or newly invested equity capital. In this chapter, I consider retained earnings. In the next chapter, I review alternative external financing sources.

∴ Company growth is the typical and desirable by-product of success of a high-technology company. Company markets, competitors, and personnel all exert considerable pressure for growth.

 # Return on Equity: Measuring Your Company's Ability to Finance Growth

Increases in retained earnings are derived from profits. The rate at which retained earnings grow is a function of the company's return on equity (ROE): net income as a percentage of total equity capital (net worth). ROE is really the fundamental measure of profitability from the viewpoint of the firm's owners, its shareholders. It is also the fundamental determinant of just how fast a company can grow by means of internal financing.

ROE defines maximum growth by internal financing.

The numerator of the ROE ratio is the company's net profit after all expenses, including interest on borrowed funds and income tax expense. It is that small portion of the firm's revenues that accrues to the benefit of the shareholders. A part or all of this net profit may be either paid out to the shareholders as dividends or retained in the firm. The denominator of the return on equity ratio is the total of shareholder investment, including both amounts the company received from the sale of common shares (and preferred shares, if any) it issued from time to time and net income earned previously and retained by the firm.

The higher the net income and the lower the amount of dividends paid, the greater the additions to retained earnings and the more this source contributes to financing future growth. Virtually all companies would prefer to finance their growth internally—that is, by means of retained earnings—rather than externally through additional borrowing or the sale of equity shares.

∴ A company's ability to sustain its growth from internal financial sources is a function of its return on equity (ROE).

➴ *Determining the Amount of Investment Your Firm Requires to Grow*

How much additional investment is required to finance a given amount of growth? The accounting equation, and thus the form of the balance sheet of a corporation (or partnership or sole proprietorship), is as follows:

Assets must grow—and be financed—as sales grow.

$$\text{Assets} = \text{Liabilities} + \text{Owners' Equity}$$

The left-hand side of this equation, the assets, represents the value of all investments in the business, from cash to accounts receivable to inventory to fixed assets and so forth. As the business grows, so will these assets. (A discussion of how these assets are valued, including the impact of inflation and fixed asset accounting, is beyond the scope of this book.)

The growth rate of assets equals the growth rate of sales, if we leave aside just for a moment the possibility that the company might become more or less efficient in utilizing those assets. Thus, we can make the reasonable *initial* assumption that, for example, a 30 percent growth in annual sales rate will necessitate a 30 percent growth in investment in assets. If a company had $7 million of assets last year when it produced $10 million of sales and this year the firm's sales grow to $13 million, a growth of 30 percent, we can reasonably expect that the company will require total assets of $9.1 million (1.3 × $7 million) to support this higher sales volume. That is, accounts receivable will expand by 30 percent if customers neither accelerate nor lag their payments next year as compared with this year. Inventory will also grow by 30 percent if the company is neither more nor less efficient in turning over its inventory. We can make similar assumptions about all other asset categories. (I return to the questions of accounts receivable collections and inventory turnover presently; the no change assumptions made here are only preliminary.)

If assets grow by 30 percent, the right-hand side of the equation (the other side of the balance sheet) must also grow by 30 percent. That is, if investments grow by 30 percent, the total sources of funds for those investments—liabilities plus owners' equity—must also grow by 30 percent. (Net worth is an alternative name for owners' equity.)

Expansion of Owners' Equity

By how much does owners' equity grow? The return on equity defines the growth rate of owners' equity *if* no new shares of stock are sold and (an important additional assumption) no dividends are paid to shareholders. That is, if the company's return on equity is 15 percent, total owners' equity at the end of the year will be, by definition, 15 percent higher than at the beginning of the year.

Now assume that assets grow by 30 percent and owners' equity grows by 15 percent. What will happen to total liabil-

ROE equals the maximum internal growth rate of owners' equity.

ities, the total debt of the business? At first glance, it might appear that liabilities will grow by (30 minus 15) 15 percent. In fact, however, liabilities will have to grow by substantially more than 15 percent in order to make up for the ROE short-fall by comparison with asset growth. The exact amount by which liabilities will have to grow is a function of the ratio between liabilities and owners' equity on the right-hand side of the balance sheet—that is, the debt leverage of the business.

The Effect of Growth on Debt Leverage

The phenomenon of debt leverage and the effect of growth on it are illustrated in Table 6-1. Case A is the condition I have posed here: The company is growing (30 percent per year) at a rate well above its ROE (15 percent per year). The result is that liabilities must grow at a substantially higher rate than

Table 6-1.   Effect of growth on the balance sheet

	End of Year	Growth Rate	Assets	Liabilities	Owners' Equity	Debt Leverage[a]
Case A:						
Company	1	—	$100	$ 50	$50	2.0
growth	2	30%	130	72.5	57.5	2.3
exceeds	3	30%	169	102.9	66.1	2.6
ROE						
of 15%						
Case B:						
Company	1	—	$100	$50	$50	2.0
growth	2	15%	115	57.5	57.5	2.0
equals	3	15%	132.2	66.1	66.1	2.0
ROE						
of 15%						
Case C:						
Company	1	—	$100	$50	$50	2.0
growth is	2	10%	110	52.5	57.5	1.9
less than	3	10%	121	54.9	66.1	1.8
ROE						
of 15%						

Note: Dividends are assumed to be $0.

[a]Assets ÷ Equity. The higher the ratio, the more the company is relying on debt to finance its growth.

assets. Here the growth rate of liabilities over the two years is 43.5 percent per year.

Case B in Table 6-1 assumes that growth of both the company's sales and its ROE are 15 percent. The result is that all elements of the balance sheet—assets, liabilities, and owners' equity—grow at the same rate, 15 percent per year; and the ratio between liabilities and owners' equity stays constant. Debt leverage neither increases nor decreases.

When growth rates exceed ROE, debt leverage increases.

Case C assumes that ROE exceeds the company's sales growth rate. The result should be predictable: Liabilities will need to grow at a slower rate than assets. Here the rate is only 4.8 percent per year. Although the liabilities are increasing, the proportions on the right-hand side of the balance sheet are changing. At the end of year 1, liabilities and owners' equity are each 50 percent of assets; at the end of year 3, liabilities are only 45 percent of assets, and owners' equity is 55 percent. Debt leverage is decreasing.

In Table 6-1, different assumptions as to the initial (year 1) proportion of liabilities and owners' equity would substantially affect the amounts of additional borrowing, the growth rate in borrowing, and the resulting proportion of liabilities and owners' equity at the end of year 3.

∴ In case A, debt leverage is increasing; in case B, it remains the same—liabilities, owners' equity, and assets are all growing apace; and in case C, debt leverage is decreasing, even though total borrowings are growing slowly. Thus, growth, ROE, and debt leverage are linked.

➡ *Three Avenues to Increased ROE*

Table 6-1 reinforces the concept that a company's ROE is a primary determinant of its ability to finance growth without resorting to either additional equity financing or excessive amounts of additional borrowing. The driving factors behind the company's ROE therefore deserve additional exploration.

Higher ROS results in higher ROE.

There are two equivalent definitions of ROE:

$$\text{Return on equity} = \frac{\text{Net Income}}{\text{Owners' Equity}}$$

$$\text{Return on equity} = \frac{\text{Net Income}}{\text{Sales}} \times \frac{\text{Sales}}{\text{Assets}} \times \frac{\text{Assets}}{\text{Owners' Equity}}$$

These equations must be equivalent, because, in the second equation, sales and assets cancel, leaving the first equation. The second, expanded, form of the ROE equation, however, permits us to focus usefully on the three key factors that dictate return on equity:

1. $\dfrac{\text{Net Income}}{\text{Sales}}$ This ratio is also called the return on sales ratio (ROS). It represents the rate of profitability per dollar of sales.

Improved asset management improves ROE.

2. $\dfrac{\text{Sales}}{\text{Assets}}$ This asset-turnover ratio indicates the productivity of business assets—that is, the dollars of annual sales that can be generated per dollar invested in assets.

3. $\dfrac{\text{Assets}}{\text{Owners' Equity}}$ This ratio is a measure of debt leverage in the business. Assets are the sum of liabilities and owners' equity; so the larger the difference between assets and owners' equity, the greater the company's borrowing and the higher this ratio.

Thus, return on equity is the product of (1) return (profitability) on sales (ROS), (2) total asset turnover, and (3) debt leverage. Increases in return on equity—a key objective for management—can be achieved in one or a combination of three ways:

1. Increasing the rate of profitability per dollar of sales
2. Increasing the efficiency with which assets are used
3. Increasing debt leverage

Of course, declines in ROE will result from the deterioration of any of these ratios. Many managers, however, would not characterize a decline in debt leverage as a deterioration. Nevertheless, reduced debt leverage will result in lower return on equity, although the attendant reduction in financing risk might make such reduction wholly appropriate and desirable.

Increased debt leverage increases ROE.

The dynamics of these interrelationships can be best illustrated by looking at a hypothetical company, first in the base case and then with improvements in ROS and in asset management (sales divided by assets).

Hitech Corporation

Figures 6-1, 6-2, and 6-3 present the financial statements of the Hitech Corporation for the years 19X0 through 19X2. In the base case shown in Figure 6-1, the company is growing at 30 percent per year. Its return on sales is 5 percent; its total asset turnover (sales divided by assets) at the end of 19X0 is about 1.5 times; and its ratio of assets to equity at December 31, 19X0, is 1.82. (These are reasonably typical ratios for U.S. companies manufacturing electronic instruments and equipment.) As a result, the company's return on (year-end) equity for 19X0 is $(.05 \times 1.5 \times 1.82 =)$ 13.7 percent. However, because in this base case the company is growing annually at a substantially higher percentage than its ROE, neither the leverage nor the return on equity remains constant in 19X1 and 19X2.

Three key assumptions in Figure 6-1 are that ROS remains at 5 percent for the following two years, total asset turnover remains at 1.5 as the company grows, and any shortfall in financial resources is made up by borrowing. Thus, the total borrowings of $1745 and $2710 shown in Figure 6-1 at December 31, 19X1, and 19X2, respectively, are derived figures: Total assets required are a function of company growth and the balance sheet is *balanced* by borrowing, as required.

Figure 6-1 illustrates the increased debt leverage that accompanies rapid growth of many high-technology companies. Note the result. Debt leverage, measured as assets divided by equity, moves from 1.82 at December 31, 19X0, to 2.01 in 19X1 to 2.18 in 19X2. The company is becoming progressively

Balance sheet ($000)		December 31	
Assets	19X0	19X1	19X2
Cash	$ 200	$ 260	$ 340
Accounts receivable	1,645	2,140	2,780
Inventory	2,000	2,600	3,380
Total current assets	3,845	5,000	6,500
Fixed assets, net	2,800	3,640	4,730
Total assets	$ 6,645	8,640	$11,230
Liabilities & owners' equity			
Accounts payable	$ 800	$ 1,040	$ 1,350
Accrued liabilities	1,200	1,560	2,030
Total current liabilities	2,000	2,600	3,380
Borrowings	1,000	1,745	2,710
Total liabilities	3,000	4,345	6,090
Owners' equity:			
Invested capital	1,000	1,000	1,000
Retained earnings	2,645	3,295	4,140
Total owners' equity	3,645	4,295	5,140
Total liabilities & O.E.	$ 6,645	$ 8,640	$11,230
		Year ended December 31,	
Income statement ($000)	19X0	19X1	19X2
Sales	$10,000	$13,000	$16,900
Net income (5%)	$ 500	$ 650	$ 845

Figure 6-1. Hitech Corp. financial statements: base case, 19X0–19X2

more highly leveraged. The positive attributes of increased debt leverage are also working: Return on year-end owners' equity grows from 13.7 percent in 19X0 to 15.1 percent the following year to 16.4 percent in 19X2.

Debt leverage cannot continue to increase forever.

Nevertheless, the steady march to higher levels of debt leverage is not a sustainable phenomenon. One possible financial strategy for Hitech Corporation, to be discussed in the next chapter, is to sell periodically newly issued shares of common stock, using the proceeds from the stock sale to repay debt. Other possible means of financing a high growth rate, besides permitting debt leverage to increase, are suggested by the expanded form of the ROE equation discussed earlier: Improve the ROS (profitability as a percentage of sales) or improve asset management (more dollars of sales per dollar

Income statement ($000)	19X0	19X1	19X2
	Year ended December 31,		
Sales	$10,000	$13,000	$16,900
Net income ($)	500	910	1,520
(%)	5%	7%	9%
Balance sheet ($000)		*December 31*	
	19X0	19X1	19X2
Assets[a]	$ 6,645	$ 8,640	$11,230
Liabilities & owners' equity			
Total current liabilities	$ 2,000	$ 2,600	$ 3,380
Borrowings	1,000	1,485	1,775
Total liabilities	3,000	4,085	5,155
Owners' equity:			
Invested capital	1,000	1,000	1,000
Retained earnings	2,645	3,555	5,075
Total owners' equity	3,645	4,555	6,075
Total liabilities & owners' equity	$ 6,645	$ 8,640	$11,230

[a]See Figure 6-1.

Figure 6-2. Hitech Corp. financial statements: improved profitability, 19X0–19X2

of assets or, equivalently, fewer dollars of assets per dollar of sales). These more palatable alternatives are illustrated in Figures 6-2 and 6-3.

Improving ROS

Improving return on sales has the most dramatic impact on ROE, whether you achieve this improvement by means of higher prices, lower product costs, or lower overhead. Of course, all companies would wish for improvement in ROS. I have in earlier chapters pointed out avenues to improved ROS—from technological superiority to process improvement to enhanced quality. Strong ROS performance—closer to 10 percent than 5 percent for most high-technology companies—is a necessary prerequisite to internal financing of a high rate of growth.

You need a strong ROS to finance internally a high growth rate.

Figure 6-2 assumes a steady improvement in return on sales from 5 percent in 19X0 to 7 percent in 19X1 to 9 percent in the final year. This added profitability causes retained earnings to grow faster than in Figure 6-1, and these increased retained earnings are substituting for borrowing to finance Hitech's growth. Assets are the same in both Figures 6-1 and 6-2, but total borrowings at the end of 19X2 are now only $1775, or nearly $1 million less than the $2710 shown in Figure 6-1. (Cumulative profit through 19X1 and 19X2 is, of course, greater in Figure 6-2 than in Figure 6-1 by the same amount, $935,000.) Although total borrowings in Figure 6-2 have grown over the two years, debt leverage (again measured by assets divided by equity) has not changed substantially: The ratio is 1.82 in 19X0, 1.90 in 19X1, and 1.85 in 19X2. The company's debt leverage has remained about constant. ROE has now increased sharply (13.7 percent in 19X0, 20.0 percent in 19X1, and 25.0 percent in 19X2) because of the increased rate of profitability on sales, not because of increased debt leverage.

Improving Asset Management

Don't overlook better asset management as a way to finance growth or reduce leverage.

An often overlooked but particularly intriguing and desirable way to improve the company's ability to finance growth is illustrated in Figure 6-3: Improve asset management. In recent years, particularly with the advent of very high interest rates and increased difficulty of accessing both the equity and long-term debt markets, managers of all companies, including high-technology companies, have been paying increasing attention to asset management—improving their balance sheet ratios. The payoff potential is very significant, as we shall see. Managers who have been traditionally income statement oriented—with a focus on earnings per share and ROS—are recognizing that reducing the investment in assets required to develop, produce, and sell the company's products or services can sharply improve the company's overall performance.

The income statement assumptions underlying Figure 6-3 are those that appear in Figure 6-1: 30 percent per year growth in sales and a steady return on sales of 5 percent. (Note that Figure 6-3 does not assume the improving rate of profitability of Figure 6-2.) The key differences between Figures 6-1 and 6-3 are summarized in the asset management assump-

Asset management assumptions			
Accounts receivable collection (days)	60	53	46
Inventory turnover (X per year)	5	6	7
Fixed asset turnover (X per year)	3.6	4.0	4.4
Total asset turnover	1.50	1.72	1.94

Income statement—see Figure 6-1

Balance sheet ($000)		*December 31*	
Assets	19X0	19X1	19X2
Cash	$ 200	$ 260	$ 340
Accounts receivable	1,645	1,890	2,130
Inventory	2,000	2,170	2,410
Total current assets	3,845	4,320	4,880
Fixed assets, net	2,800	3,250	3,840
Total assets	$ 6,645	$ 7,570	$ 8,720
Liabilities & owners' equity			
Total current liabilities	$ 2,000	$ 2,600	$ 3,380
Borrowings	1,000	675	200
Total owners' equity	3,645	4,295	5,140
Total liabilities & owners' equity	$ 6,645	$ 7,570	$ 8,720

Figure 6-3. Hitech Corp. financial statements: improved asset management, 19X0–19X2

tions shown at the top of Figure 6-3: The accounts receivable collection period improves from sixty days in 19X0 to fifty-three days and finally to forty-six days in 19X2; inventory turnover (sales divided by inventory) improves from five to six and finally to seven times per year; and fixed asset turnover (sales divided by fixed assets) also evidences steady improvement, from 3.6 to 4.0 to 4.4. (The better managed high-technology companies have demonstrated that improvements in asset utilization of about these magnitudes are fully achievable.) The result: Total asset turnover—one of the key factors in the expanded version of the ROE formula—increases from 1.5 to 1.72 in 19X1 to 1.94 in 19X2. Corollary results are as follows:

1. Total borrowings actually decrease over the two-year period in spite of the 30 percent compound annual growth rate and no improvement in ROS.

Reduce asset intensity to ease the task of financing high growth.

2. Debt leverage (assets divided by equity) declines from 1.82 in 19X0 to 1.76 to 1.70. Although the absolute change in this ratio is not great, the change in trend—from increasing in Figure 6-1 to decreasing here—is very significant, given the assumed high growth rate.

3. ROE improves slightly, by comparison with Figure 6-1, but the improvement from 13.7 percent to 16.4 percent is achieved not with increased, but with decreased, debt leverage, or financing risk.

In actual fact, these favorable results are understated. ROE would increase more and ROS would have improved somewhat in comparison with Figure 6.1. In these simplified presentations, savings in interest expense that accompany debt reduction are ignored in the calculation of profit (additions to retained earnings).

Such are the mathematics of improving ROE through enhanced asset utilization and thereby bettering the company's ability to finance its growth. Increasing asset turnover is easier said than done. Not only is concerted management effort required, but efficient asset management must be accorded high priority—parallel with or above that given to meeting the company's shipment goals and near-term net earnings targets. Without this increased focus on assets, the organization will tend to overinvest in assets to be sure to meet shipment and profit targets. How does this overinvestment occur? Inventories and accounts receivable are the prime offenders in high-technology companies.

Inventory.   When inventory planners focus on the income statement and not assets, the motto becomes: "When in doubt, order extra." Raw and in-process inventories are allowed to balloon to minimize the possibility of delaying a shipment or disrupting in-process manufacturing. Finished goods inventories are maintained at a high level so that customer orders can be filled immediately, thus enhancing current period earnings, even if the customers themselves would tolerate some wait.

The inventory and production planning staff is not the only culprit. As discussed earlier, careful inventory planning requires the cooperation of all functions of the business. Marketing must commit to a shipment schedule with sufficient lead time so that manufacturing can plan orderly production flow and minimize safety stocks that are otherwise required to meet last minute changes in customer demand. Engineering documentation must be sufficiently complete and accurate—and engineering designs frozen in a timely manner—so that sophisticated inventory and production planning schemes, including MRP systems, can operate effectively. Procurement must be imaginative in contracting with suppliers for blanket order agreements that will permit frequent deliveries of small quantities, as Japanese manufacturers have done as part of their "just in time" planning systems. Accounting and finance must conform to invoice payment terms negotiated with these key suppliers on whom the company depends. Both manufacturing and engineering must be committed to a level of quality that will reduce the need for safety stocks, smooth the production flow, and reduce the inventory of replacement parts for warranty repair. The culprit for excessive inventory investment is generally assumed to be manufacturing; but I have stressed throughout Chapters 2 through 5 that the problems that manifest themselves in high inventories are generally multifunctional.

Reducing inventory should be a companywide priority.

Spare parts inventories are particularly troublesome for suppliers of sophisticated equipment and systems, who frequently find that they must maintain substantial inventories of replacement parts. A spare parts depot should typically be maintained separate from production inventories (so inventory withdrawals to satisfy the random demand for spare parts do not interrupt the smooth flow of production), and a number of field depots will frequently be required as well. At the extreme, when systems are large and downtime is very costly to customers, spare parts inventories must be maintained at customer sites.

Spare parts inventories are particularly costly.

Programs to reduce investments in service inventories should include one or all of the following steps:

1. Improve system reliability and/or improve the effectiveness of preventive maintenance in order to reduce the need for emergency parts shipments.

2. Increase the use of air freight or other means of expedited delivery to reduce the need for redundant inventories. (In comparing the cost of air freight and the cost of carrying inventory, guard against underestimating the inventory carrying cost, which can easily run to 30 or 40 percent per year.)

3. Educate customers to accept slower response time for emergency repair or to invest in their own inventory of emergency repair parts.

4. Standardize on components or modules across product lines, or reduce the frequency of model changes, so that a given investment in spare parts can serve a broader set of customer installations.

5. Reduce the number of years over which the company guarantees to maintain parts and service for its older product lines. When this period has passed, eliminate spare parts being inventoried for this obsolete product.

Accounts receivable are usually the largest asset investment.

Accounts Receivable. Too frequently, accounts receivable collection periods are assumed to be beyond the company's control, or at least out of the hands of other than the company's credit department. Slow-paying customers may be a manifestation of deeper and broader problems, and solving these problems can yield benefits in addition to freeing up funds formerly invested in accounts receivable.

The credit department is the first line of defense against "stretched" accounts receivable. When management places insufficient emphasis on asset management, the credit department has limited power to intercede in the planned sale of products or services to customers with weak credit ratings. The pressure for sales and near-term profits will rule the day, and the result may be a long delay in collecting the receivable (or its ultimate write-off as a bad debt).

A 25 percent speedup in collections can often reduce total assets by 10 percent or more.

Speeding up collections can drastically affect the company's total asset base and thus its ability to finance its growth internally. In many high-technology companies, particularly those requiring few fixed assets, accounts receivable may account for 40 percent of total assets in the business. In the case of service companies requiring no inventories, this percentage may be 60 or 70 percent. A reduction from sixty to forty-five days in the average collection period for a company whose accounts receivable are 40 percent of assets represents a 10 percent reduction in total asset investment—a worthy goal.

In periods of tight money and high interest rates, all customers are motivated to delay invoice payments unless or until such practices disrupt relationships with present or future suppliers. Thus, the credit department's job is not limited to preapproving a proposed sale. It must pursue routine and timely follow-up of outstanding invoices. The "squeaking wheel" syndrome suggests that the supplier who is actively (but politely!) demanding timely payment of outstanding invoices is likely to gain prior claim on any available cash of a customer feeling the pinch of tight money. Providing cash discounts to customers in return for prompt payment is another device to accelerate collections, but one that is not widely used in high-technology industries. Even a discount as high as 2 percent for payment within ten days of the invoice date may not stimulate early payment. If a customer, in the absence of taking the discount, will pay in sixty days, he or she is in essence borrowing for fifty days (sixty less ten) for 2 percent, an annualized interest rate of only about 14.6 percent. The customer's alternative may be to borrow from a bank at rates that, at least in recent years, have frequently exceeded 15 percent.

But the problems are generally deeper than tight money or weak credit. Slow-paying customers may in fact be unhappy customers. Customers who are dissatisfied with the quality of the components received, rejecting large numbers of shipments, and customers who find equipment or systems to be unreliable or service to be unresponsive are understandably going to be tardy in paying their bills. Moreover, purchasers of high-technology capital equipment, who must rely on the manufacturer for both installation and personnel training, will delay approving invoices for payment until installation and training have been completed and start-up has been successfully accomplished. Other buyers of finished products (as contrasted with components) who receive incomplete shipments—portions of the product or system having been back ordered by the supplier—will justifiably delay payment until

Slow-paying customers are often unhappy customers.

the shipment is complete and the product's capabilities thoroughly tested.

A concerted attack on extended accounts receivable requires the attention and cooperation of all functions of the business. Thus, when stretched receivables plague the company, a task force that includes representatives from the following functions, in addition to those from the credit department, should attack the problem:

Attack extended accounts receivable with a task force from the credit department, marketing, manufacturing, field service, training, accounting, and engineering.

1. Marketing. In addition to ferreting out systemic problems in the order-receipt-delivery-billing cycle, individual accounts must be scrutinized and the following types of questions asked: Does the customer dispute the bill? Were promises of performance beyond normal specification made—that is, was the customer oversold? Was the customer urged (or coerced) to accept early delivery so the salesperson could meet his or her quota or earn an extra bonus?

2. Manufacturing: Is the number of back orders excessive? Are unacceptable delays encountered in filling them? Is the quality level adequate and within the specifications agreed to with the customer?

3. Field service: Was installation accomplished on schedule and were operating specifications met? Is the customer withholding payment to pressure the service department to perform?

4. Training: Have up-to-date instructions and manuals been supplied? Has training been accomplished on schedule?

5. Accounting: Are invoices prepared and mailed with all deliberate speed? Are they accurate? Do they conform to the order acknowledgment? Does the customer dispute the bill?

6. Engineering: If custom-engineered products or services were supplied, were agreed-upon specifications met? Is delivery complete? Is promised documentation complete?

General managers should view any deterioration in the aging of accounts receivable—that is, a slowdown in collections—as an important early warning signal of customer dissatisfaction, which probably arises from one or more of the problems previously enumerated.

I emphasized shipment nonlinearity—that is, rushed shipments at the end of an accounting period—in Chapter 3 because of its disruptive effects on the relationship between the marketing and production departments in both inventories and accounts receivable. In-process inventories are built to distressingly high levels just prior to period end, and safety stocks of components and subassemblies are maintained at extra high levels to facilitate the last-minute rush of shipments. Of course, accounts receivable peaks at the end of the period. The high-technology company must finance both of these peaks.

Nonlinear shipments demand unnecessarily high investments in accounts receivable.

Table 6-2 illustrates the effect of the end-of-period syndrome on the investment in accounts receivable. Both companies A and B ship $300 per fiscal quarter and have an accounts receivable collection period of forty-five days. However, company A achieves linear shipments, and company B makes two-thirds of its quarterly shipments in the final month of the quarter. At the end of the quarter, company B's accounts receivable balance is $225, as compared to $150 for company A. Thus, although the average investment level in accounts receivable is the same for the two companies ($150), the peak investments are very different. Company B has a peak 50 percent greater than company A. The peak, not the average, dic-

Table 6-2. Effect of end-of-period syndrome on accounts receivable investment

Shipment Pattern

	Month			Quarter Total
	1	2	3	
Co. A	$100	$100	$100	$300
Co. B	$ 50	$ 50	$200	$300

Month-End Investment in Accounts Receivable
(Assuming a 45-Day Collection Period)

	Month			Average
	1	2	3	
Co. A	$150	$150	$150	$150
Co. B	$150	$ 75	$225	$150

tates the level of required investment. Company B must arrange its financial structure to accommodate this peak investment.

Fixed Assets.   Although accounts receivable and inventory—that is, working capital—are the key undermanaged assets in high-technology companies, fixed asset investments deserve some additional mention. Most high-technology managers exercise substantially more care, and perform a good deal more analysis, with respect to fixed asset investments than for working capital investments because the former are typically premeditated and are represented by a relatively few, large acquisitions. In contrast, overinvestment in accounts receivable and inventory result from many small decisions: Purchase additional components for inventory; add another part or subassembly to the spare parts stock; sell one order to a customer with marginal credit; or ship one system incomplete or without adequate final testing.

A company's requirements for fixed assets are very much a function of its integration strategy, as well as its policies regarding capacity (see Chapters 3 and 4). A company that elects to backward integrate must typically invest substantially more in fixed assets than a company that chooses to rely on subcontractors. A company aggressively pursuing the benefits of the experience curve phenomenon—that is, scale economies—typically pursues a strategy that results in heavy investment in fixed assets. A company that elects to build additional capacity in anticipation of demand requires earlier investments in fixed assets, and lower fixed asset turnovers, than a company whose capacity additions lag demand. These investments are inherently neither wise nor unwise. But the firm's manufacturing and financial strategies must be *integrated* to assure mutual consistency.

Integrate your firm's financial and manufacturing strategies.

It is characteristic of high-technology companies that attractive investment opportunities far outstrip available financial resources. Thus, capital budgeting becomes a process of ranking investment alternatives to assure that those offering the highest returns are assigned the highest priority. (I discuss some problems encountered in applying conventional capital budgeting techniques in a high-technology business in the next section of this chapter.) In such an environment, investments in product or market extensions—that is, investments in technological development and market growth—are likely to prove more attractive than fixed asset investments, particularly investments in plant and office space. Not surprisingly, therefore, emerging high-technology firms wisely seek to decrease their fixed asset intensity. Two popular steps are to lease rather than own facilities and to rely on subcontractors.

Leasing facilities and subcontracting production reduce asset intensity.

However, we must not ignore the relative riskiness of alternative investment opportunities. A low-risk, lower-return investment in fixed assets may well prove to be a more judicious investment for a high-technology company than a high-risk, high-return investment in, for example, a state-of-the-art product development project.

∴ Improving return on sales improves return on equity, a prerequisite to internally financing a high growth rate. Better asset management is a particularly desirable way to free up funds internally to finance growth. Accounts receivable and inventories are typically the dominant assets in a high-technology company. Improving collection periods and inventory turnover is the responsibility of all functions of the business, including marketing, field service, training, manufacturing, and accounting.

➡ *Analyzing Capital Investments in High-Technology Companies*

Throughout this chapter, I have stressed the benefits of reducing your firm's asset investments to increase financial returns, reduce debt leverage, and improve the company's ability to finance its growth internally. The very essence of general management is resource allocation—deciding what investments to make, when, and where. The objective is efficient and profitable investment, certainly not an absolute minimum of investment.

Some of the accepted methods of investment analysis—capital budgeting—encounter real shortcomings in dealing with high-risk investments, the type that characterize high-risk, high-technology companies. Discounted cash flow techniques (often referred to as DCF/ROI), which have come into common use in recent years, have been blamed by some people for inadequate investments in risky, but potentially very profitable new technology.[1] In fact, these analytical techniques are as theoretically sound for evaluating a proposed investment in developing a new technology, product, or market as for evaluating a proposed investment in a machine tool or in real estate (the types of investments to which they are more usually applied). However, incomplete or faulty application of the techniques is widespread. Such misapplication can have the unfortunate effect of discouraging investment in high-risk, but high-payoff, development projects.

Don't avoid high-risk investments.

High-technology companies must make high-risk investments if they are to remain at the leading edge of technology and competitive in their marketplace. Avoiding high-risk investments, and concentrating the capital budget on "safe" investments, will soon turn a high-technology company into a low-technology company. But high-technology companies need not discard accepted analytical techniques. Rather, they need to guard against the excessive conservatism that tends to get factored into investment analyses.

Six key problems are commonly encountered in applying DCF/ROI evaluation techniques to high-risk development projects. First, payoffs from these projects are typically highly skewed, and the mean or modal (or most likely) values understate the expected value, calculated probabilistically. Analysts are likely to utilize the modal values of future cash inflows in their analyses, and these modal values give no recognition to the finite (although perhaps small) probability that the development project will result in a very high payoff. These exceptionally high payoffs can occur when the advent of a revolutionary new product causes a new market, whose

Don't ignore the small probability of a very high payoff.

dimensions far exceed normal expectations, to emerge. Recall the very low initial market projections for revolutionary products such as the mainframe computer, the office copier, and instant (one-step) photography.

Second, analysts are likely to assign high "risk adjustments" to the future cash flows from development projects. These adjustments may be implemented by increasing the return that is required (increasing the hurdle rate) or decreasing (implicitly or explicitly) the future cash inflows or—too frequently—both. The magnitude of the risk adjustments is likely to be geared to the perceived risk of the first stage of the investment, the product development phase. However, subsequent investments in market development and in productive capacity will be made only if the initial product development activity is successful.

Once the first stage of the investment has been completed successfully, the riskiness of the balance of the project is substantially reduced. As a result, although early cash outflows can reasonably be subjected to high risk adjustments, subsequent outflows and inflows should not suffer the same penalty. Maintaining high risk adjustments throughout the multistage, multiyear project penalizes the development project inappropriately.

Use different risk adjustments for different stages of the investment.

Third, and related to the first two factors, is the typical failure to recognize explicitly when analyzing a complex, multiyear development project that management will face opportunities both to cut losses, if early phases of the development prove unpromising, or to accelerate and amplify investments, if early development uncovers unforeseen opportunities.

Thus, the cash outflows associated with the development can be truncated when and if management discerns that the probability of failure is overwhelming, thereby reducing the cost of the "failure" outcome. Conversely, if highly favorable outcomes are realized—those outcomes to which low probabilities had been assigned in project evaluation—further investments can be undertaken to realize the full potential of the development and thereby increase the entire project's payoff.

Losses can be cut and gains can be amplified.

Fourth, development projects are generally better protected than investments in productive capacity from the ravages of inflation. The future cash inflows and outflows that will result from project acceptance are likely to track inflation or, in economic jargon, fully adjust for inflation. Investments in fixed assets, by contrast, will not fully adjust, in major part because the future tax shield associated with depreciation is a function of today's values and does not inflate.

Don't ignore inflation's effects on future net cash inflows.

Analysts too frequently ignore the effects of inflation on future net cash inflows while discounting them at rates geared to today's costs of capital, costs that may reflect the capital markets' expectations of high inflation. Development projects, when evaluated on a DCF/ROI basis, will be penalized unless future cash flows are inflated before they are discounted at a required rate of return that implicitly reflects high inflation or real (uninflated) cash flows are discounted at a required rate of return that reflects real capital costs (that is, capital costs without the inflation premium).

Take advantage of the portfolio effect.

Fifth, the high risk associated with any single project is mitigated by the diversity of projects the company undertakes. This effect is referred to in financial theory as the portfolio effect. If the company is at any time pursuing a variety of projects involving somewhat different technologies and aimed at somewhat different markets and if no single project is so large that its failure would prove catastrophic to the company, the company has diversified its development risk, just as an individual investor might diversify his or her portfolio of common stock investments. The risk associated with the total portfolio of projects—that is, with the entire development budget—is substantially less than that associated with any single project. Risk adjustments on individual projects should therefore be mitigated to reflect this lower risk associated with the portfolio of projects.

Don't compound risk adjustments.

Sixth, both analysts and managers have a tendency to overestimate development projects' riskiness and thus to make excessive risk adjustments. This phenomenon is particularly prevalent in large companies, where middle managers perceive the personal risk associated with backing an unsuccessful investment as substantially higher than the personal rewards that attend successful risk taking. Projects initiated and approved at lower management levels are subjected to an extensive review process, with the possibility that additional risk adjustments will be imposed upon them. Compounded risk adjustments may result in ultimate rejection.

Projects screened out at lower management levels are not typically subjected to any further reviews that might bring to light excessive risk adjustments. Excessive risk adjustment will result in rejecting development projects that, if accepted, would enhance the overall wealth of company shareholders.

Declining rates of R&D spending in the United States and the related problems of low gains in productivity and lagging innovation are almost surely due in part to the misapplication—or incomplete application—of the DCF/ROI technique to the analyses of investments in development projects. Risks must be taken, not avoided (or even minimized), in capital budgeting in high-technology companies.

Do not use DCF/ROI techniques blindly.

∴ DCF/ROI, if applied correctly, is as valid for analyzing high-risk investments as it is for analyzing any other kind. If high-technology companies avoid high-risk investments, they soon become low-technology companies. The primary problems encountered in applying DCF/ROI techniques in high-technology companies relate to excessive risk adjustments, failure to allow fully for very high payoff possibilities, and improper analysis of inflationary effects.

Highlights

- The high-technology company's markets, competitors, and personnel exert considerable pressure to grow at rates well above the average for all industries.

- Return on equity is *the* measure of a company's ability to finance its growth internally—that is, from profits.

- ROE is the product of ROS, debt leverage, and asset turnover.

- Improvements in ROS can have the most dramatic effect on ROE.

- In a high-technology company, accounts receivable and inventories are typically the dominant assets.

- Improvements in inventory turnover require the concerted effort of manufacturing, marketing, and engineering.

- Stretched accounts receivable can be symptomatic of deeper problems.

- A task force consisting of persons from marketing, manufacturing, field service, training, accounting, engineering, and the credit department should be used to attack the problem of stretched receivables.

- Risky investments must be accepted and not avoided in high-technology companies.

- Capital budgeting in high-technology companies must avoid the problems of excessive risk adjustments and allow fully for possible very high payoffs and for inflation.

Notes

1. See, for example, Eugene L. Grant, W. Grant Ireson, and Richard S. Leavenworth, *Engineering Economy*, 7th ed. (New York: Wiley, 1981).

Chapter Seven

Financing High-Technology Companies

This chapter covers the following topics:

- How to Determine the Appropriate Debt Leverage: Exploring the Trade-Offs
- Sources of Debt Financing
- Equity Capital: The Fundamental Source of Financing
- Financing Strategies of Some Successful High-Technology Companies
- Long-Range Financial Planning: Modeling Financial Statements
- Evolution of the Financial Function as a High-Technology Company Grows

We turn now from internal sources of financing to external sources: primarily debt and common-stock equity. In the context of considering these alternative sources of capital, I analyze capital structures of high-technology companies. Capital structure—that is, how the business is financed—both influences and is influenced by the capital intensity of the business. The appropriate capital structure is a function of the other risks, including the technological risk, in the business.

In reviewing financing alternatives in this chapter, we look at the optimum level of debt leverage; debt financing (including short-term, long-term, and lease financing, which can in most cases be viewed as a form of intermediate-term debt financing); the equity markets—from the venture capital market (a capital market devoted in a major way to high-technology industries) to public underwritings and "being acquired" as a special case of equity financing; some examples of alternative financing strategies that successful high-technology companies have pursued; the useful procedure of attempting to reflect the economics of a high-technology company in a model of its financial statements so as to consider explicitly the impact of changes in strategy on the firm's returns, asset intensity, and capital structure; and the evolution of the financial function within a high-technology company as it grows from a start-up to a middle-sized company.

How to Determine the Appropriate Debt Leverage: Exploring the Trade-Offs

Trade off risk of failure and high returns.

All companies utilize debt financing. They are at least indebted to the telephone company from the time of a long-distance call to the time the telephone bill is paid and to their employees between a particular workday and the following payday. At most, they finance virtually all their assets with debt. The issue, then, is one of degree: To what extent should the company rely on borrowing rather than shareholder investment to finance its growth? That is, what is the correct amount of debt leverage for a high-technology company? Financial theorists argue endlessly over this question. There is no single correct answer; every high-technology company must trade off between risk and return.

In Chapter 6, I noted that an increased (or leveraged) return on equity is accompanied by an increased risk of wide swings in profit and possible failure of the company. Borrowing reduces the need for equity capital, thus permitting, under favorable profit conditions, improved returns for the shareholders. But borrowing also carries with it the absolute promise to pay interest and principal when due; inability to make these payments may result in corporate bankruptcy.

The fundamental determinant of "appropriate" debt leverage is the shareholders' risk preferences. (As a practical matter, however, management's risk preferences typically dominate.) Do the shareholders and management want to "eat well" or "sleep well"? That is, where do they strike the balance between the pursuit of profit leverage and the avoidance of risk? Managers differ widely. Some borrow all they can, allowing debt leverage to increase to the maximum amount the credit sources will tolerate. Others have an aversion to borrowing, opting instead for the lower growth or lower equity returns—but also lower risk—that attend reduced borrowing.

The more stable your company's cash flow, the greater debt leverage you can afford.

In general, the more stable a company's cash flow, the more debt leverage it should employ. The extreme is a public utility, for example, an electric power company. State governments grant monopoly positions to utilities. They can count on steady customer demand for their product and are permitted by regulatory authorities to price so as to earn an agreed-upon return. The utility's cash flow is therefore steady and predictable, and it is reasonably assured that it can service— pay interest and principal on—its borrowings. Thus, the utility can (and should) employ a good deal of debt in its capital structure.

High-technology companies are typically at the other end of the stability spectrum from public utilities. Technology is changing rapidly; customer demand is unpredictable; competitive environments alter radically and quickly; and price erosion may be dramatic. Cash flow is anything but steady and predictable. In such an environment, to rely heavily on debt financing is to compound the company's risks: The financing risk is added to the not inconsequential technological, market, and competitive risks the company faces. Credit sources are cognizant of this potential compounding of risks. High levels of debt are often simply unavailable to the high-technology company.

Generally, opt for low debt leverage in high-technology companies.

As a general rule, high-technology companies should opt for below-average debt leverage. In terms of asset-equity ratios—the measure of debt leverage introduced in Chapter 6—financial plans should typically aim at a ratio below about 1.8, and those managements more averse to financial risk (perhaps because of high market or technological risk) should seek to hold the ratio below 1.5. But these targets should be adjusted for individual companies, bearing the following in mind:

1. Larger companies, serving several markets with a range of products based on a number of technologies, can tolerate higher asset-equity ratios than can smaller companies with a narrow and more vulnerable product range.

2. More fixed asset–intensive businesses can utilize long-term debt (and lease) financing and can therefore appropriately pursue higher debt leverage.

3. A smaller company growing very rapidly may undertake periodic new equity financings. Between these financings, it relies on debt; thus, its assets-equity ratio cycles just as does its financing.

"Surprises" almost always increase debt leverage.

Untoward events may push up debt leverage well beyond planned levels. Indeed, the frequency with which these events occur in technology-based companies argues for planning on a conservative (low) debt leverage position. The following are examples of such events:

1. Planned additions to retained earnings fail to materialize or, worse, operating losses erode company equity.

2. Asset investments climb sharply, particularly inventory (when sales slow, new models are introduced, or operating problems forestall the completion of in-process inventory) and accounts receivable (when quality and performance problems in the field cause customers to delay payments).

3. A planned equity offering has to be postponed or canceled at the last minute because of sudden shifts in investment climate or near-term financial results for the company that are below investor expectations.

When setting the company's debt leverage strategy, recall Murphy's Law: "What can go wrong, will." And whatever goes wrong will almost always increase debt and may also reduce equity. Asset-equity ratios can accelerate both radically and suddenly. Consider the well-publicized fates of Lockheed Corporation and Memorex in the 1970s, International Harvester in the early 1980s, and a host of smaller, less-visible companies in the 1982–1983 recession.

∴ External sources of financing are borrowing and newly issued equity securities. The proper mix of these two sources—the appropriate debt leverage—is a function of the asset intensity of the business; the stability of its cash flow; market, technology, and other risks; and management's and shareholders' risk preference.

Sources of Debt Financing

All companies use some short-term debt financing: amounts due to vendors, employees, and the tax collectors, at a minimum. In addition to these sources of debt capital, there are short-term borrowing, long-term borrowing, and lease financing (a kind of quasi debt).

Short-Term Borrowing

Commercial banks are the primary source of interest-bearing short-term debt, but other current liabilities should not be overlooked as sources of company funding; some are noninterest bearing.

Accounts receivable are the best collateral for a short-term bank loan.

Bank Borrowing. Most bank lending to high-technology companies is secured, directly or indirectly. That is, the bank seeks as collateral (security) for its loan those company assets it could attach in the event of default. A high-technology company's best collateral is its accounts receivable; so borrowing against pledged receivables is an enormously important source of bank financing for high-technology companies.

Bankers focus on two key questions as they consider the value of accounts receivable as collateral:

1. What is the credit worthiness of the typical customer? A company selling technology products to Fortune 500 companies is able to borrow more readily on its accounts receivable than is a company selling to undercapitalized retail stores specializing in personal computers.
2. Does the product work and will the customer keep it? In the early stages of a high-technology company, before customers have accepted the product, initial sales may represent little more than "consignment of inventory." That is, the risk of customer returns is high. The bank is counting on the receipt of cash, not the return of merchandise, to settle the customer's account receivable.

Banks want to know if your customers are credit worthy and if your product works.

Assuming customers of average credit worthiness and products that are well beyond the prototype stage, a high-technology company should be able to negotiate a short-term line of credit with its commercial bank that will permit it to borrow up to the equivalent of about 80 percent of so-called *eligible accounts receivable*. Eligible receivables are typically defined as receivables that are not past due (for example, not more than sixty or ninety days from invoice date) and are with customers whose credit is acceptable to the bank. Customers are normally not notified of this "pledging" of receivables, and customer payments come to the company rather than to the bank. Sometimes banks lend against specific customer receivables (when each receivable is very large). In other cases, they simply require monthly verification that aggregate receivables provide sufficient collateral for the amount borrowed under the line of credit.

Inventory is substantially less attractive than accounts receivable as a source of collateral. The more specialized the inventory, the less acceptable it is to the bank. Work-in-process inventory—that is, semifinished inventory—is practically valueless as collateral. Should the company fail, the bank is in no position to complete and recover on partially completed products. However, some electronics companies may have large inventories of gold, copper, or silver. These raw materials are easily attached and readily sold by the bank and thus represent outstanding collateral. When inventory consists largely of standard components, the bank typically lends up to 50 percent of the value of that inventory, in addition to any amounts for which accounts receivable serve as collateral.

Banks may accept some inventories or orders as collateral.

Some engineering development companies are able to borrow against firm contracts from reputable customers—that is, borrow against an order before it turns into an account receivable. If the company's contracts call for "best efforts" development, rather than delivery of a developed product to meet guaranteed performance specifications, and if the company has a strong reputation and good working relationships with its customer (for example, the federal government), it should press its commercial bank to accept these contracts as collateral on about the same basis (50 percent) as inventory.

Early stage technology companies often find that banks "request" the personal guarantee of the dominant owner/managers as additional collateral for the company's borrowings. The personal guarantee of a wealthy individual is particularly beneficial to the bank, but the bank may even seek a guarantee from an entrepreneur who has no wealth outside his or her business. The purpose of such a guarantee is to ensure that the entrepreneur will not "walk away from" the business in the event of severe financial trouble. Entrepreneurs and owner/managers should weigh carefully whether they are willing to expose their personal assets to the risk that attends a personal guarantee. Sometimes they have no practical choice, but personal guarantees are a legitimate point of negotiation between the borrower and the bank.

Consider carefully before offering your personal guarantee for your company's loan.

Virtually all short-term bank borrowing involves important provisions in addition to the identification of collateral. Normally, certain minimum liquidity and maximum debt leverage ratios must be maintained. The bank may restrict payment of salaries and dividends and investment in fixed assets in order to assure that the borrowed funds are used for the purposes agreed to. Various other covenants are included in the agreement, and default under any covenant causes the loan to be immediately due and payable. It is not unusual, however, for bankers to waive one, or even several, violations of such loan covenants, particularly if a demand for immediate loan repayment would force the company into bankruptcy (or a distress sale or merger). Nevertheless, even technical default of a short-term loan provides the bank with substantial power over the affairs of the company and its management, an uncomfortable situation whether or not the bank chooses to exercise that power. Thus borrowers should thoroughly understand and carefully review all covenants. Negotiations should center on just a few critical covenants, particularly minimum working capital and profit requirements.

Focus negotiations on key loan covenants.

Traditionally, commercial banks have viewed short-term credit as seasonal; the borrower is expected to "clean up" the borrowing (reduce it to zero) at some point during the year. Most high-technology companies experience little seasonality and yet they are substantial short-term borrowers. In these cases, although the maturity on their loans may be ninety or a hundred eighty days, both the bankers and the borrowers expect the loans will simply be renewed at maturity. Such short-term loans that do not clean up at some time during the year are often referred to as *evergreen loans*. Even evergreen loans must ultimately be repaid, from either company profits or the proceeds from the sale of equity securities. Banks are reluctant to get locked into evergreen loans unless one or both of these sources of repayment appear viable.

Don't overlook stretched accounts payable as a funding source.

Other Current Liabilities.   The other current liability that represents the most important source of company funding is accounts payable. Within limits, it represents an essentially free source of short-term funds. A company whose accounts payable balance is $1 million when it pays its vendors in thirty days (on average) will have an accounts payable balance of $1.5 million—and an additional $500,000 of capital to invest— if it delays payment to vendors to an average of forty-five days. Some cash discounts may be forgone in connection with this stretching of accounts payable, but these lost discounts often entail substantially less cost (or opportunity loss) to the company than borrowing the equivalent amount of money from the bank, assuming that the bank is even willing to lend it.

Some cautions should be heeded regarding this source of funds, however. When faced with unreasonable stretching of payments by a customer, most alert suppliers will take one or a combination of the following actions:

1. Hold up new shipments to extract payment of past-due invoices. Such delays can wreak havoc with production schedules, leading to missed deliveries (and delays in cash receipts from customers), excessive overtime expense, and so forth.

Stretched payments can hamper purchasing's negotiations.

2. Refuse to accept additional orders. When the supplier knows that its customer depends upon a critical component or device for its high-technology product or system, this threat of breaking off the business relationship can be effective in extracting prompt payment. This tactic is particularly likely when the supplier is operating at or above total capacity.

3. Offer less-attractive prices. A company that develops a "slow pay" reputation handicaps its purchasing department in its efforts to negotiate favorable prices. The linkage between price penalties and stretched accounts payable can be hard to discern, but, again, alert suppliers and alert purchasing agents recognize that a trade-off exists between prices and payment terms.

4. Provide less-favored treatment to the customer. One of the dimensions on which a customer relationship is valued is promptness of payment. Again, the linkages may be hazy, but they are real. The most valued customers will be accorded readier access to the suppliers' engineers for advice and will be kept abreast of technological developments, changes in specifications, new product plans, and so forth. More-favored customers are likely to be given delivery priority over less-favored customers. Most suppliers keep careful track of just which customers are to be accorded most favored treatment.

Notwithstanding these four possible problems, accounts payable should not be overlooked as an important source of financing. A key to availing your company of extended payment terms without undue penalty is to be open and honest with your suppliers: Include payment terms as part of your purchasing negotiations.

Include payment terms as part of your purchasing negotiations.

Just as delayed payments to suppliers can represent a source of funds, so can accelerated payments from customers—that is, customer down payments. High-technology capital equipment is frequently sold on terms that provide for up to 25 percent payment at the time the order is placed. The more specialized (custom) the equipment, the more likely that down payments will develop as normal terms of sale within the industry. For large, one-of-a-kind systems, suppliers may be able to demand progress payments as well as a down payment; so as much as 90 percent of the cash may be received from the customer by the date of delivery.

The dramatic effect that down payments can have on your company's need for working capital is illustrated in Figure 7-1. To the extent that you get down payments, you create a current liability (source of funds). This current liability (customer down payments) is a low-risk liability because it does

Assumptions for table below:

1. Accounts receivable are collected from customers 45 days following shipment date (A/R period).
2. Down payments are received from customers 90 days prior to shipment date (DP period).
3. Shipments are steady at $1 per day.

Down Payment Percentage	Current Asset: Accounts Receivable Balance	Current Liability: Customer Down Payments	Difference	
			$	% of Monthly Sales
-0-	$45.0	-0-	$45.0	150%
5%	42.8	$ 4.5	38.3	128%
10%	40.5	9.0	31.5	105%
15%	38.3	13.5	24.8	83%
20%	36.0	18.0	18.0	60%
25%	33.8	22.5	11.3	38%
30%	31.5	27.0	4.5	15%
35%	29.3	31.5	(2.2)	(7%)

	DP Period	A/R Period
A	90 days	45 days
B	90 days	30 days
C	90 days	60 days
X	60 days	45 days
Y	60 days	30 days
Z	60 days	60 days

Figure 7-1. Impact of down payment terms on working capital requirements

not represent a direct call on the firm's cash flow. The liability is discharged by performance, rather than by the payment of cash.

In addition, down payments reduce the amount remaining to be collected once shipment is effected. Thus, the *net* working capital required for financing sales to customers is really the difference between the reduced accounts receivable balance and the down payment current liability.

Figure 7-1 indicates that, under its restrictive but realistic set of assumptions, this element of working capital investment is ten times greater if no down payment is obtained by comparison with the situation where a 30 percent down payment is obtained. A 10 percent down payment reduces (by comparison with zero down payment) this element of working capital by nearly one-third, and a 25 percent down payment reduces it by about 75 percent.

Customer down payments dramatically reduce working capital requirements.

Customers resist making down payments (for the same reasons that they are attractive to suppliers), thus making down payment terms a frequent competitive pricing weapon. Weaker or newer entrants to the market may attempt to gain a competitive advantage by reducing or eliminating down payments; in many industries, the result has been the disappearance of the practice of demanding down payments.

Long-Term Borrowing

Borrowings with maturities of greater than one year are long-term borrowings. Such debt of a more permanent nature is available only to the more credit worthy of high-technology companies. Frequently these borrowings are secured, with the collateral being fixed assets or long-term accounts receivable that arise from installment sales or leases to customers.

Bank term loans typically have maturities between two and five years and provide either for amortization during the term of the loan or for a balloon payment of part or all of the principal at maturity. As with short-term bank borrowing, restrictive covenants impose some loss of flexibility on management, for example, in the payment of dividends or the acquisition and sale of business units.

It is difficult to generalize about either the availability of long-term bank borrowing or the terms of such borrowing when it is available. It is clear, however, that banks are reluctant to enter term loan agreements with high-technology companies until those companies have demonstrated success in their markets and achieved a healthy profitability record. Even then interest rates (generally higher than rates on short-term borrowing), maturity (typically from three to five years, but sometimes as long as seven or eight), collateral (both current and noncurrent assets are possible), and covenants vary widely from company to company and bank to bank. Some shopping around among banks is frequently a good idea; the "shopping," however, must be done with finesse, not as if you were buying a used car, because banks seek "lending relationships," not just borrowers.

Don't be reluctant to shop discreetly among banks.

When a high-technology company needs multiyear financing (that is, the need is neither seasonal nor transitory) and can negotiate reasonable terms (including interest rate), it should seek a term loan rather than an evergreen line of credit. There are two primary reasons. First, the bank's lending commitment is locked in for the term of the loan, while a line of credit is callable every ninety or one hundred eighty days. Second, total credit available to the company is likely to be expanded if both term and line of credit borrowing agreements are in place. This extra credit may never be needed, but its availability reduces financing risks to the business.

Equity Kickers.   Loans with longer terms, of up to, say, fifteen years, may be available to larger, stable technology-based companies. The source of these loans is typically insurance companies and pension funds, rather than commercial banks. As the term extends, lenders are likely to demand what are commonly referred to as *equity kickers:* an opportunity to share in the expected future appreciation of the company's common stock. Lenders argue that, because these loans provide long-term, permanent capital for the business, in a sense, they displace shareholder investment. Put another way, the alternate source for the high-technology company is the sale of additional shares of common or preferred stock. Because these

Insurance companies and pension funds may require equity kickers.

lenders accept the additional risk associated with longer-term loans, they seek compensation in the form of participation in the equity of the business.

The two most common forms of equity kickers are convertible debt and debt with warrants. Table 7-1 illustrates these financing vehicles that are so widely used by emerging, high-growth technical companies.

In convertible debt, the loan itself is convertible into common shares of the company. That is, at the lender's election, the loan can be exchanged for a prenegotiated number of shares of common stock (in Table 7-1, $2 million ÷ $30 per share = 66,667 shares). The conversion price is typically set above the market price that prevails at the time of loan negotiation. In this case, the conversion price, $30, is 20 percent above the market price, $25. The interest rate on the debt is set below prevailing seven-year term interest rates. The rate here is 8 percent, and the corresponding rate on a seven-year note without conversion privileges might be 14 percent. A trade-off exists between conversion price premium and interest rate: The higher that premium, the less the lender will gain from market appreciation of the common stock and the more the lender will charge in interest. As noted in Table 7-1, the average yield to the lender can be very attractive if the market value of the borrower's common stock appreciates significantly.

Equity kickers can result in lower interest rates and larger loans.

When warrants accompany the debt, the loan agreement provides the lender with warrants (options) to purchase a predetermined number of shares at a predetermined price, and the loan is not convertible. Table 7-1 calls for warrants to

Table 7-1. Illustration of equity kickers: $2 million, 7-year loan[a]

	Interest Rate	Conversion or Exercise Price[b]	Number of Warrants	Assumed Market Price at Conversion/ Exercise (After 5 Years)[c]	Effective Pretax Yield/Cost[d]
Convertible debt	8%	$ 30	N/A	$60	21% per year
Debt with warrants	11%	$ 30	20,000	$60	15.5% per year

a. Assuming no amortization of principal.
b. Market price at time of negotiation = $25.
c. Assumes market price appreciation of 19.1% per year.
d. Including both interest and capital appreciation.

purchase twenty thousand shares at $30 per share. (The exercise value of these warrants is 20,000 shares × $30 per share = $600,000.) Thus, this lender has access to only 30 percent of the amount of equity available under the terms of the convertible debt. The "warrant coverage" is said to be 30 percent. As a result, the interest rate on the debt is higher—11 percent as compared with 8 percent—although still below the rate that would prevail in the absence of any equity kicker.

In both cases, the lender will not exercise its right to obtain equity—either by conversion or the exercise of the warrants—unless (or until) the market price of the borrower's common stock exceeds by a comfortable margin the conversion or warrant price, in this illustration, $30. If the borrower enjoys explosive and profitable growth and this success is reflected in appreciation of its common stock price, the lender may reap substantial profits from these loans—profits that translate into yields well in excess of the prevailing 14 percent interest rate on equivalent "straight" debt.

Loans with equity kickers are a compromise between straight debt and the sale of common shares.

The borrower benefits in several ways. First, the interest rate on the debt is lowered by the existence of the equity kickers. Second, the alternative source of capital is typically the sale of common stock at $25 per share (the prevailing market price at the time of negotiation). If, instead, the convertible loan is negotiated and, after several years, the lender converts it into common stock, the company will have, in effect, sold its common stock for $30 per share—the conversion price—rather than $25 per share. Finally, such loans are typically subordinated to the senior debt of the firm. That is, the lender, in return for the equity kicker, agrees to take a junior position to other short- or long-term borrowing; should the company fail, senior lenders will be repaid first. Subordination may, in turn, open up opportunities for the company to obtain still more bank borrowing because the bank will view the subordinated borrowing as somewhat equivalent to equity. Nevertheless, subordinated loans are still absolute obligations of the company; failure to pay the interest and principal on the subordinated debt when due represents default, just as it does with senior debt.

Long-term debt with equity kickers has proven to be a very important source of financing for smaller technology-based companies. The substantial appreciation in common stock prices of many high-technology companies over the past twenty-five years or so has caused many lenders to be attracted to this form of lending, which provides both down-side protection and up-side potential. If the company does not enjoy great success, the lender retains the protection of the debt instrument (with its several covenants and default provisions). If the company prospers, the lender shares in that prosperity through the exercise of either the warrants or the conversion privilege. *Equity kickers provide lenders down-side protection and up-side potential.*

Because of the fickleness of the credit markets, subordinated debt with equity kickers is not always available. When it is, and when your high-technology company qualifies for this specialized form of financing, it is a very attractive compromise between straight term borrowing (at high interest rates) and the sale of common stock (at low per-share prices). One of the qualifying conditions, incidentally, is a good likelihood that a public market will exist for the company's common stock when conversion is effected or the warrants are exercised.

Deferred Income Taxes.   Deferred income taxes, another long-term liability and therefore source of funds, arise because of a difference in timing between when revenues or expenses are recognized for financial accounting (book or shareholder reporting) and when they are recognized for tax reporting purposes. In most instances, the tax deferral is essentially permanent; thus, this liability takes on many of the characteristics of equity. Better than both debt and equity, deferred taxes neither obligate the company to pay interest nor dilute the shareholders' interest.

Managers should always seek both to minimize and postpone paying income taxes (within the context of tax laws and regulations). Deferred income taxes arise frequently and can represent a substantial source of funds. High-technology companies should use the following opportunities to defer income taxes: *Always try to minimize and postpone paying income taxes.*

1. Accelerate depreciation for tax purposes even if straight-line depreciation is used for book purposes (A high-tech-

nology company whose operations are not fixed asset-intensive will not derive much benefit here.)

2. Report sales and finance leases on the installment basis for tax purposes (recognizing revenue and the associated cost of sales pro rata as cash is received over the life of the contract), and report them at "full value" for book purposes at the time of shipment

Offsetting these tax deferrals, unfortunately, are certain prepaid taxes that arise primarily from reserves accrued for book purposes and disallowed for tax purposes. In high-technology companies, three such reserves are often both important and large—those for future warranty expenditures, obsolete inventory, and customer bad debts. High-technology companies need to marshal their arguments to convince the income tax auditors that substantial additions to these reserves are necessary and thus should be deductible, given technological and credit risks.

Lease Financing

Use leases as a form of intermediate-term borrowing.

Leasing is an important source of intermediate-term financing. Many high-technology firms lease both their facilities and much of their machinery and equipment. These leases, particularly for assets other than real estate, are typically full-payout finance leases—that is, they are in most respects the equivalent of intermediate-term borrowing. "Full payout" means that the monthly lease payments over the lease term will aggregate to the full purchase price of the asset plus interest. Stated another way, the risks and rewards of ownership of the asset pass to the lessee (the user), and the lessor (the owner) acts simply as the financing agent.

What are the motivations for leasing rather than borrowing and owning the asset? First, most high-technology companies should deploy their scarce and expensive financial resources in higher-risk, higher-priority investments—for example, new product development—rather than in buildings or machinery. In effect, leasing can expand the total amount of "borrowed" capital available to the company, and lease financing is typically less expensive than equity financing. In circumstances where a bank might be unwilling to lend the money required for a smaller high-technology company to purchase, for example, a computer or a small fleet of forklift trucks, a lessor may be attracted. The lessor is better able than the bank to repossess the leased assets—and re-lease them to others—should the lessee fall on hard times.

Invest scarce resources in high-return projects, and lease your buildings and machinery.

Another advantage of leasing is that the tax benefits associated with fixed asset ownership can flow to the lessor—to whom they are more valuable—rather than to the lessee. For example, an embryonic high-technology company facing several years of operating losses before significant profitability should lease rather than own assets. If it owns the assets, it must postpone the benefits of accelerated depreciation and investment tax credits that accompany fixed asset investments until the company becomes profitable. Alternatively, the lessor can take advantage of the tax benefits of ownership immediately, reflecting these benefits in part in a lower lease rate to the lessee.

Historically, smaller companies have sought to lease rather than borrow and own assets in order to dress up their balance sheets: The lease obligation, although fixed and absolute, would not appear on the company's balance sheets; thus, the company's *apparent* debt leverage would be reduced. However, in recent years, the accounting profession has insisted that such leases be reflected on the balance sheet for what they are: the equivalent of borrowing. Thus, balance sheet dressing no longer represents an important motivation for leasing.

Don't count on leases to dress up your balance sheet.

∴ Short-term debt financing is obtained largely from commercial banks, which in turn seek collateral (often accounts receivable) and protective loan covenants. Because customer down payments can both reduce accounts receivable balances (a use of funds) and increase current liabilities (a source of funds), they are an attractive form of short-term borrowing when competitive conditions permit their use in standard terms of sale. Long-term debt financing of medium-sized high-technology companies typically requires equity kickers. Leasing represents an important source of financing that is equivalent to secured, intermediate-term borrowing.

Equity Capital: The Fundamental Source of Financing

In addition to retained earnings and borrowed capital, the sale of newly issued shares of common stock is an important source of financing. This is the fundamental and first source of company financing: All corporations must have shares of common stock outstanding.

There are four primary markets for equity capital for high-technology companies, aside from funds that may be invested by the founding group: venture capital market, private placement market, public market, and company acquisition market. Although the boundary line between these markets is often fuzzy, I consider each separately. I have already discussed a form of quasi-equity financing: term debt with equity kickers. This form of security may appear more like debt or more like equity, depending upon the relative importance the lender and borrower place upon the interest rate of the debt and the size and terms of the equity kicker. A convertible debt that carries a minimal interest rate and provides for zero conversion price premium is in essence an equity financing, regardless of how it is labeled.

Venture Capital Market

The venture capital market is segmented by source of funds and stage and type of investment.

The U.S. venture capital market is an amazing phenomenon that is marvelously peculiar to this country. Over the past twenty years, it has grown into a reasonably mature and orderly market of substantial size. It began with the investment activities of certain wealthy family trusts, particularly the Rockefellers, the Phipps, and the Whitneys. As the market has expanded, it has also segmented. Venture capital comes now from a wide variety of sources: wealthy individuals, major (typically mature) industrial companies, commercial banks, investment banks, insurance companies, pension funds, and various endowment funds. The investment decisions themselves are most frequently made now by the general partners of venture capital partnerships that have as limited partners the institutions just enumerated. In addition, some commercial and investment bankers, as well as a few industrial companies, operate their own venture capital funds.

Some venture capital funds specialize by areas of technology, concentrating their investments in, for example, computer and related products, integrated-circuit manufacturers, software firms, or bioengineering companies. Others specialize not so much by industry as by investment stage, focusing on start-up companies or second- and subsequent-round financings. These later stage financings of high-technology companies occur after development risks have passed and when the companies are gearing up for large scale production and expanded marketing.

Although certain venture capital firms concentrate on nontechnology opportunities (for example, consumer service firms or leveraged buy-outs of business units from large companies), high-technology industries have captured the attention of most of the traditional venture capital firms. The high-growth opportunities, coupled with low capital intensity and the emergence of a series of niche markets in which strong competitive positions can be built, have provided spectacular payoffs on certain venture capital investments. These success stories, together with a 1978 change in the U.S. capital gains tax laws, have attracted increasing amounts of money to the venture capital market. During the four years following the reduction in the capital gains tax rate in 1978, several billion dollars came into the venture capital market. Interestingly, the number of start-up firms seems to have kept pace with the increased availability of venture capital investment funds.

High-growth technology-based firms capture most venture capitalists' attention.

Cost. A first concern of many potential seekers of venture capital is the reputed high cost of this form of financing. Venture capitalists who invest in start-up companies seek to earn on the order of ten times their investment within five years, which translates to a compound rate of return of almost 60 percent. Early investments in companies such as Rolm, Intel, and Genentech have resulted in substantially higher returns.

Venture capitalists seek big payoffs— understandably.

But all start-up investments do not yield such heady returns. These spectacular successes compensate for the majority of venture investments, which turn out to be mediocre or worse, including 10 to 20 percent that ultimately end up in bankruptcy.

Selling additional common stock later in its life cycle is less costly to the emerging company. As the early stage technological and management risks dissolve, the price of a share of common stock in the company increases, and investors at the second or third round may foresee the opportunity for a compounded rate of return of only 20, 30, or 40 percent, rather than 60 percent.

Guidelines. The seeker of venture capital financing for his or her start-up or emerging high-technology company should bear in mind the following suggestions when negotiating for venture capital:

Use these guidelines when negotiating for venture capital.

1. Venture capitalists place primary emphasis on the management team's quality, integrity, ability, and experience. They seek complete and well-rounded management teams. A first-rate team pursuing a second-rate product-market opportunity is more likely to obtain venture capital financing than is a second-rate team pursuing a first-rate idea.

2. The preferred opportunity allows the high-technology company to develop and hold a commanding market share position in a niche market. A small market share position in a large market is unlikely to be highly profitable, for reasons discussed in earlier chapters.

3. Venture capitalists seek liquidity for their investment within five to seven years. The practical sources of liquidity are a public offering of the company's common stock or the acquisition of the company by a larger firm.

4. Venture capitalists expect the firm to be professionally managed, with traditional economic objectives. They shy away from companies that are to be operated primarily for the scientific or intellectual amusement of the principal managers and engineers. Venture capitalists expect managers to seek rapid growth and to be motivated to build a major enterprise.

5. Venture capital is not a common source of funds for fundamental research. Engineering activities during the early years of the company should be much more heavily weighted toward development than toward research. (However, a number of companies in recent years have used tax-oriented R&D limited partnerships to finance large, high-risk research and development projects.)

6. Venture capitalists seek very high returns on their investments, but they do not seek to "steal" the entrepreneurs' ideas or to control the business. Venture capitalists are investors, not managers. They expect the company to be operated by management; they expect the founders to be the management unless or until they prove incapable of the task, in which case the venture capitalists want to be able to replace them. Most venture capitalists limit their roles in the company; they seek only to serve on the company's board of directors and to act as informal advisors to management. However, bear in mind that venture capital arms of mature industrial companies often seek windows on new technology that may be useful to their parent companies or association with a young venture that might later be an attractive acquisition candidate.

Venture capitalists seek high returns, competent managers, and unique product and market niches.

7. The majority of venture capitalists are honorable and can be relied upon to maintain the confidentiality of the entrepreneurs' ideas.

8. Venture capitalists want to be certain that the entrepreneurial team has very high incentive for success. As a result, the entrepreneurial team should have sufficient ownership in the company to maintain very high motivation. In recent years, it has not been unusual for venture capitalists to invest $2 million or more in a company in return for, say, 65 percent of the equity, permitting the remaining 35 percent to go to the entrepreneurs, who are investing very little cash.

9. Venture capitalists expect each member of the founding team to invest a significant portion of his or her net worth in the new company—enough to heighten motivation and to cause a failure to "hurt" but not devastate the founder. If a member has no net worth (or conceivably a negative net worth), some venture capitalists may even lend the founder money to invest in the start-up company.

*Use a business
plan and shop
discreetly for
venture capital.*

10. The venture capital community is tight-knit. Directories of venture capital firms are available.[1] With a bit of homework and reference checking, you can develop a short list of appropriate potential investors. Initiate contacts with only a few investor groups at the outset. You can expect the firms you contact to be in communication with each other.

11. Prepare a written business plan, with a succinct executive summary at the outset. Venture capitalists view the business plan as a starting point in their investigation. If interested, they will request a great deal of additional information. The written business plan permits venture capitalists to assess the founding team's ability to reason, communicate, plan, and anticipate problems.

12. Some venture capitalists prefer straight equity investment, but, more typically, they invest in convertible preferred stock or convertible debt, with the objective of placing their investment in a preferred position vis-à-vis the common stock owned by the management. Such a capital structure also permits management to achieve significant ownership position with minimum capital investment and without unfavorable tax consequences.

13. Significant legal pitfalls surround the offering and sale of securities in a firm, whether they are to be sold to professional venture capitalists or others. Competent legal and tax advice is essential for both the entrepreneurial team and the venture capitalists.

Private Placement Market

The private placement market is an institutional market, dominated by insurance companies, pension funds, endowment funds, and, increasingly, foreign capital. Once a high-technology company has gained a certain degree of success—and eliminated certain risks—these more traditional and less expensive sources of equity (or quasi-equity) capital are available to it. This market is particularly attractive to the young company that is not yet ready, or able, to tap the public market for new equity funds but offers a risk-return profile that no longer requires great venturesomeness on the part of the investor. Even publicly owned companies may find the private placement market attractive when conditions either in the public market or within the firm preclude the sale of additional shares of stock in a public underwriting at a reasonable price.

The private placement market is midway between the venture capital and the public markets.

The venture capital market is a private placement market of a particular type and the demarcation between the venture capital and other segments of the private placement markets is becoming somewhat less clear. Recently, several venture capital partnerships have raised substantial amounts (for example, over $100 million) from their limited partners, in part to permit the partnerships to make later stage, large investments in the potentially more successful of their companies.

Foreign investors and industrial companies often participate in private placements for smaller technology-based firms in order to gain *increased knowledge* of the technology or market the smaller company is pursuing, place themselves in a preferred position for a possible *subsequent acquisition* of all the shares of the technology-based company, and realize *capital appreciation*, as would any other investor.

The cost to the issuing company of equity capital sold in the private placement market is typically lower than the cost of venture capital financing, but higher than the cost of equity funds raised in a public underwriting. Private placement investors, like the venture capitalists, seek ultimate liquidity for their investment, probably within two or three years. Investment advisors—for example, investment bankers or corporate finance departments of commercial banks—should typically be retained by those high-technology companies seeking access to the private placement market. Fees to these advisors, although not modest, are usually contingent upon successful completion of the financing.

Public Market

The public market is usually available only to proven high-technology companies.

Some very small technology-based companies—and even some brand new ventures—seek and obtain equity financing from a public underwriting of their securities, but typically this source of capital is (and probably should be) available only to the high-technology company that has operated for several years and has achieved a respectable level of sales and profits. The threshold size for a company before it can realistically look to a public underwriting of its common stock is on the order of $20 million in annual sales and at least $1 million in annual after-tax profits. Several factors dictate this threshold:

1. The more competent, reliable, and prestigious of the underwriting firms are unwilling to assume the risks associated with underwriting the offering of a high-technology company with a short or spotty track record.
2. A substantial amount of funds—at least $5 million—needs to be raised in order to
   - Spread the high fixed costs of the underwriting (legal, accounting, printing and management time) over a substantial amount of capital so that the transaction costs are not a disproportionate percentage of the capital raised
   - Provide the major underwriting firm, and its affiliated firms, with a sufficient number of shares to command the attention of large brokerage sales forces
   - Ensure that the number and aggregate value of the shares offered are high enough to create active trading in the so-called "aftermarket" (that is, the continuing buying and selling of shares among the public once the underwriting has been completed)
3. The burdens of the underwriting itself, and of the subsequent legal and moral obligations that attend public ownership of the firm's securities, can overpower both the management and the financial resources of a very small firm.

Liquidity, access to more lower-cost financing, and ego gratification are the primary advantages of "being public."

Advantages. When access to the public market is available and appropriate, it typically represents a substantially lower-cost source of funds than does venture capital or the private placement market. Once the company is public, investors—including any venture capitalists who may have backed the company initially, as well as the company's employees—gain liquidity: They can sell stock from time to time to realize on their investment (subject to certain legal restrictions as to both the timing and the amount of the sales). Incentive stock options to management (discussed in Chapter 8) also take on more meaning when a public market has been established for the company's securities. Moreover, once the company is publicly owned, it is in a better position to acquire other companies, using its stock as the medium of payment. It can also more easily return to the public market in the future for additional equity financing. The initial successful underwriting of the company's securities also can represent a gratifying milestone of achievement for management.

Disadvantages.   The tangible and intangible costs of being a publicly owned company are less obvious. Some of the key potential negatives are the following:

1. The costs of the initial underwriting (fixed costs totaling $250,000 or more are not uncommon, even for a small offering) and of the subsequent reporting to shareholders and the Securities and Exchange Commission are not insubstantial.

   The public market is costly and tyrannical, and liquidity is often illusory.

2. The management and directors are exposed to liabilities that arise from the initial underwriting of, and the subsequent trading in, the securities. The federal and state securities laws are complex; violating them can lead to punishments that seem out of proportion to the offense.

3. The tyranny of quarterly financial reporting may restrict management's flexibility. Once public, many companies find that they are unwilling to make decisions that have favorable long-term, but unfavorable short-term, consequences. The private company is less concerned with steady progression of quarterly earnings.

4. The liquidity that investors and managers seek may be illusory. If the volume of share trading in the aftermarket is light (the market is "thin"), the purchase or sale of more than several hundred shares may cause a sharp and unfortunate movement in the stock's market price.

5. As a result, the stock's market price may be a reasonably poor indicator of true value. This problem is particularly severe on the down side. If interest in the company's stock evaporates, the market price may become—and remain—depressed. The consequences are multifold and all unhappy: Options become worthless and management becomes demoralized; shareholders' liquidity disappears; future

underwritings of the company's common stock can occur only at prices that are pegged to the depressed prices quoted in the public market; and the securities become relatively worthless as the medium of exchange in an acquisition of another company. The value of a publicly owned company, then, is dictated by the share price quoted in the daily *Wall Street Journal*. By contrast, the value of a privately owned company is a matter of negotiation between the buyer and seller of the securities.

The timing of both initial and subsequent underwritings of the common stock of smaller technology-based companies is all important. Fashions or fads seem to overtake the public market from time to time, and when the fashion is running in favor of the particular technology-based company, incredibly high prices can be obtained for company shares. Timing in the initial sale of the securities of Apple Computer and of Genentech was coincident with great public and investor interest in personal computers and genetic engineering. As a result, immediately following the underwriting, each company's share price soared well above the offering price because demand far exceeded supply among stock traders. (Subsequently, by the way, the prices of both securities settled back to levels near the original offering prices.) There are numerous examples of companies that sought public underwritings of their securities when the public market had no appetite. The typical result is that the company withdraws the planned underwriting and seeks capital from other sources, waiting for better timing for that public underwriting.

Timing is critical in going public.

A key message here is simply to strike when the iron is hot rather than delaying in hopes of obtaining a slightly higher market price. A corollary message is to maintain sufficient financial flexibility so that your company does not become absolutely dependent upon doing a public offering at a particular time.

Acquisition Market

Not to be overlooked as a source of additional capital to fuel the growth of a high-technology firm is the acquisition market—that is, the sale of all or a substantial portion of the company to another firm. A sale of securities that results in the larger company acquiring only a minority interest (that is, not acquiring control) represents, in effect, the private placement of equity securities by the smaller company with the larger company. But when a sale of common stock results in the transfer of ownership control to the acquiring company, the price realized is typically (at least in recent years) well above the price that can be realized in either the private placement or the public markets.

Prices of high-technology companies in the acquisition market are high.

The demand for technology among larger companies has bid up the prices for acquired firms. Larger companies in this country have been relatively less able than entrepreneurial ventures to develop and exploit new technology and to build entirely new markets. Thus, larger companies have found that the acquisition of smaller high-technology companies, even at startlingly high prices, represents the most efficient and effective route both to new technology and to diversification.

When should a high-techology company seek to be acquired? The decision is—and should be—based largely upon personal considerations of the top management team. I can only outline the following advantages and pitfalls that should be considered in that decision.

The sellout of a technology-based company to a larger firm often appears to be more attractive than in practice it turns out to be. The advantages are clear:

1. Investment liquidity
2. Reduction in investment risk for the present owners
3. Access to substantial financial resources and expert advice (sometimes unsought and unwanted) to accelerate the growth of the acquired company
4. Synergy arising from captive markets, compatible development activities, use of established and effective marketing channels, and so forth

Study the subtle disadvantages as well as the obvious advantages before being acquired.

The disadvantages are less clear, particularly during the courtship between the two companies. Management styles are likely to be somewhat incompatible, and decisions the acquired company used to make quickly and independently now must be cleared with the acquiring company's senior management staff. Financial reporting typically is more elaborate and complex—and costly. Corporate staff personnel may prove more meddlesome than helpful to the acquired company. Moreover, the anticipated synergy is often difficult to realize because middle management may resist adaptations in operating practices that appear to top management to offer obvious advantages.

Changes in management generally follow soon after acquisition.

In a large percentage of the cases senior managers (particularly the president) of the acquired company leave after an initial honeymoon period or at the end of their employment contract period. (Such contracts are a typical feature of these acquisitions.) The transition from the presidency of a small or medium-sized company to the general managership of a division of a larger, typically more structured, company is not easy. However, in many cases, the resulting change in top management—a change that might not occur if the company remained independent—is both inevitable and desirable. Many managers who are highly effective during the entrepreneurial phase of a company's life prove to be substantially less effective in a middle-sized company, where delegation, matrix management, and effective control assume greater importance. Thus, the top management shuffling that frequently follows the acquisition should not necessarily be taken as evidence that the acquisition was inappropriate or poorly executed. Departing senior managers frequently "recycle" themselves into entrepreneurial ventures, where their talents are particularly useful.

∴ The four sources of equity financing are the venture capital market, other private placement markets, the public market, and the acquisition market. Venture capital is a well-developed, expensive market offering many advantages to young high-technology companies. The cost of equity in the private placement market is generally somewhere between that in the venture capital and the public markets. The public market, accessible only to proven high-technology companies, provides advantages but also tangible and intangible costs to the company. The acquisition market frequently offers the highest return to shareholders.

Financing Strategies of Some Successful High-Technology Companies

There is no one correct financing strategy for high-technology companies, but there are some common themes. Figures 7-2 through 7-8 contain abbreviated financial statements for the Hewlett-Packard Company, Syntex Corporation, International Business Machines, Texas Instruments, Tandem Computers, California Microwave, and Nippon Electric Company. The ratios at the bottom of each of these figures highlight their respective financial strategies, which I describe further in the following paragraphs. Note in each case the relationships among company growth rate, financial leverage, asset intensity, profitability, and sources of outside capital.

There is no single correct financial strategy.

Hewlett-Packard is a highly successful manufacturer of electronic instruments and computing systems. By policy, HP avoids long-term debt and limits its growth rate to a level it can achieve without benefit of significant debt leverage. The company's growth rate exceeds by a considerable margin its return on equity; infusions of new equity are obtained primarily from the company's employee stock purchase plan. Thus, the stock purchase plan is both a valued employee benefit and a key element of the company's financing strategy.

HP uses an employee stock purchase plan to help finance high growth.

A manufacturer of pharmaceutical and related high-technology medical products and devices, *Syntex* spends a large percentage of its revenue on development and necessarily enjoys a high gross margin (between 60 percent and 65 percent) in order to fund heavy development and marketing expenses. The company's ROS is very strong (about 13 percent), partly as a result of a very low tax rate. This ROS mitigates the company's heavy investment in assets (total turnover less than 1.0). In spite of only moderate leverage, the company's ROE is in the 17 percent to 20 percent range.

Syntex achieves high ROS to finance heavy asset investments and realize high ROE.

IBM, the world's leading producer of computers and office products, is a giant company that has grown at a somewhat modest rate in recent years. Because much of its output is

IBM's healthy gross margin permits strong ROE with very low asset turnover.

	Years ending October 31		
	1981	1980	1979
Income statement			
($ millions)			
Sales	$3,578	$3,140	$2,527
Gross margin	1,875	1,665	1,421
Net income	312	269	203
Balance sheet			
($ millions)			
Net working capital	$1,001	$ 800	
Fixed assets	979	789	
Total assets	2,758	2,337	
Long-term liabilities	26	29	
Owners' equity (net worth)[a]	2,028	1,617	
Ratios			
Gross margin (%)	52.4	53.0	56.2
Return on sales (ROS) (%)	8.7	8.6	8.0
Working capital turnover	3.6	3.9	
Fixed asset turnover	3.7	4.0	
Total asset turnover	1.3	1.3	
Leverage (assets/ equity)	1.4	1.4	
Return on equity (ROE) (%)	15.4	16.6	

[a]Includes deferred income.

Figure 7-2. Abbreviated financial statements and ratios for Hewlett-Packard Company

leased, the company has a low asset turnover (a major part of its assets is comprised of computer systems on lease to customers) and high gross margin. Heavy development and marketing expenses (over 30 percent of revenues) are made possible by the company's healthy gross margin, permitting IBM to achieve a strong ROE in spite of low asset turnover and only moderate leverage.

	Years ending July 31		
	1981	1980	1979
Income statement *($ millions)*			
Sales	$ 711	$ 580	$ 471
Gross margin	461	363	290
Net income	99	75	64
Balance sheet *($ millions)*			
Net working capital	$ 241	$ 248	
Fixed assets	270	219	
Total assets	771	708	
Long-term liabilities	82	108	
Owners' equity (net worth)[a]	501	425	
Ratios			
Gross margin (%)	64.8	62.6	61.6
Return on sales (ROS) (%)	13.9	12.9	13.6
Working capital turnover	3.0	2.3	
Fixed asset turnover	2.6	2.7	
Total asset turnover	0.9	0.8	
Leverage (assets/ equity)	1.5	1.7	
Return on equity (ROE) (%)	19.8	17.6	

[a]Includes deferred income.

Figure 7-3 Abbreviated financial statements and ratios for
Syntex Corp.

Texas Instruments presents an interesting contrast to the
other companies, particularly HP. TI and HP are about the
same size and their ROEs are similar (except in 1981, when
TI experienced a slowdown and a drop in profit). TI's gross
margin is much below HP's—only 28 percent. Yet its high
asset turnover (particularly in working capital) and somewhat
higher debt leverage permit it to earn a respectable ROE in

*TI balances its low
margins with high
asset turnover and
debt leverage.*

	Years ending December 31		
	1981	1980	1979
Income statement			
($ millions)			
Sales, rentals, & services	$29,070	$26,213	$22,863
Gross margin	17,054	16,064	14,450
Net income	3,308	3,562	3,011
Balance sheet			
($ millions)			
Net working capital	$ 2,983	$ 3,405	
Fixed assets	7,688	6,634	
Total assets	29,586	26,703	
Long-term liabilities	3,853	3,542	
Owners' equity (net worth)[a]	18,161	16,453	
Ratios			
Gross margin (%)	58.7	61.3	63.2
Return on sales (ROS) (%)	11.4	13.6	13.2
Working capital turnover	9.7	7.7	
Fixed asset turnover	3.8	4.0	
Total asset turnover	1.0	1.0	
Leverage (assets/ equity)	1.6	1.6	
Return on equity (ROE) (%)	18.2	21.6	

[a]Includes deferred income.

Figure 7-4. Abbreviated financial statements and ratios for International Business Machines Corp. (IBM)

spite of a relatively low ROS (about 5 percent, compared to HP's of almost 9 percent).

Tandem finances explosive growth through sales of stock.

Tandem Computers was founded in 1974 and has experienced very rapid growth; its sales nearly doubled in each of the last two years. Gross margins and ROS have been strong. The company has financed growth through the sale of common stock; note the substantial increase in owners' equity in

	Years ending December 31		
	1981	1980	1979
Income statement			
($ millions)			
Sales	$4,206	$4,075	$3,224
Gross margin	967	1,152	912
Net income	109	212	173
Balance sheet			
($ millions)			
Net working capital	$ 432	$ 328	
Fixed assets	1,106	1,097	
Total assets	2,311	2,414	
Long-term liabilities	221	224	
Owners' equity (net worth)[a]	1,325	1,218	
Ratios			
Gross margin (%)	23.0	28.3	28.3
Return on sales (ROS) (%)	2.6	5.2	5.4
Working capital turnover	9.7	12.4	
Fixed asset turnover	3.8	3.7	
Total asset turnover	1.8	1.7	
Leverage (assets/ equity)	1.7	2.0	
Return on equity (ROE) (%)	8.2	17.4	

[a]Includes deferred income.

Figure 7-5. Abbreviated financial statements and ratios for Texas Instruments, Inc. (TI)

1981. Thus, Tandem avoids debt leverage probably because management perceives that technology risks and operating leverage are high. The company's ROE is on the low side (mid teens) for a "glamour" technology company. Working capital and total asset turnovers appear low in 1981 because of large cash-equivalent investments (following a public offering of the company's stock) that are available to finance the anticipated rapid growth of the firm.

	Years ending September 30		
	1981	1980	1979
Income statement			
($ millions)			
Sales	$ 208	$ 109	$ 56
Gross margin	132	68	35
Net income	27	11	5
Balance sheet			
($ millions)			
Net working capital	$ 179	$ 62	
Fixed assets	36	14	
Total assets	256	96	
Long-term liabilities	2	2	
Owners' equity (net worth)[a]	205	70	
Ratios			
Gross margin (%)	63.5	62.3	62.5
Return on sales (ROS) (%)	13.0	10.1	8.9
Working capital turnover	1.2	1.8	
Fixed asset turnover	5.8	7.8	
Total asset turnover	0.8	1.1	
Leverage (assets/ equity)	1.3	1.4	
Return on equity (ROE) (%)	13.2	15.7	

[a]Includes deferred income.

Figure 7-6. Abbreviated financial statements and ratios for Tandem Computers, Inc.

Cal Microwave's high fixed asset turnover offsets low margins.

California Microwave manufactures communication equipment and systems for which the Defense Department is a major customer. ROS is low, as it is for most defense contractors, but asset turnover is high—approaching 2.0 in the most recent year. Fixed asset turnover is by far the highest of the companies in this sample. Debt leverage is average; thus, the respectable ROE is achieved by rapid turnover of assets and in spite of low ROS.

	Years ending June 30		
	1982	1981	1980
Income statement			
($ millions)			
Sales	$ 89	$ 57	$ 38
Gross margin	19	14	10
Net income	4	3	—
Balance sheet			
($ millions)			
Net working capital	$ 16	$ 13	
Fixed assets	8	7	
Total assets	47	37	
Long-term liabilities	—	1	
Owners' equity (net worth)[a]	28	23	
Ratios			
Gross margin (%)	21.4	23.7	25.2
Return on sales (ROS) (%)	4.7	4.4	—
Working capital turnover	5.6	4.4	
Fixed asset turnover	11.1	7.7	
Total asset turnover	1.9	1.5	
Leverage (assets/ equity)	1.7	1.6	
Return on equity (ROE) (%)	14.8	10.8	

[a]Includes deferred income.

Figure 7-7. Abbreviated financial statements and ratios for California Microwave, Inc.

Nippon Electric (NEC), one of the major, diversified electronics firms in Japan, provides a startling contrast and confirms that Japanese high-technology companies earn sharply lower ROS, are more asset intensive, and are more highly leveraged than their U.S. counterparts. NEC's leverage ratio is three times that of TI, the next most highly leveraged company in this sample. ROS is generally below 2 percent and the gross margin is also low. Asset turnover is below 1.0, although

NEC relies on very high debt leverage to finance its growth.

| | Years ending March 31 | | |
	1982	1981	1980
Income statement			
(billions of yen)[a]			
Sales	1,280	1,075	879
Gross margin	392	335	272
Net income	28	22	15
Balance sheet			
(billions of yen)			
Net working capital	44	42	
Fixed assets	296	229	
Total assets	1,389	1,154	
Long-term liabilities	343	275	
Owners' equity (net worth)[b]	209	174	
Ratios			
Gross margin (%)	30.6	31.2	30.9
Return on sales (ROS) (%)	2.1	1.7	1.0
Working capital turnover	29.1	25.6	
Fixed asset turnover	4.3	4.7	
Total asset turnover	0.9	0.9	
Leverage (assets/equity)	6.7	6.6	
Return on equity (ROE) (%)	13.3	12.7	

[a]Because our attention is solely on ratios, it does not matter whether the statements are in dollars or yen.
[b]Includes deferred income.

Figure 7-8. Abbreviated financial statements and ratios for Nippon Electric Company (NEC)

working capital turnover is surprisingly high, a condition that arises not from tight control of current assets but from the fact that short-term borrowing and trade payables are major sources of financing for the firm. With an asset-equity ratio approaching 7, the company relies heavily on all forms of debt. Thus, an ROE not much below the other companies is achieved by means of debt leverage and in spite of ROS and asset turn-over ratios that appear very low.

∴ Successful high-technology companies utilize a variety of financing structures and strategies, as shown by the dis-cussion of the seven firms in this section.

 # *Long-Range Financial Planning: Modeling Financial Statements*

These illustrations of contrasting business and financial strat-egies point up the need for each high-technology company to think through its own business strategy and the effects on the company's financial statements. The pieces of the puz-zle must fit together. The financial managers—and, indeed, the whole management team—must keep the company's plans for growth rate, returns, asset management practices, and financial leverage from becoming mutually incompatible. The company needs a long-range financial plan. It must do more than compare next year's budget with this year's actual, strive for only small increases in asset turnover rates, and worry about financing only next year's growth. It must consider the fundamental nature of its business and its financial implica-tions. In short, it must take the long view.

Be explicit and realistic in planning returns, turnover, leverage, and growth rates for your firm.

As a part of this long-term financial planning process, the high-growth, high-technology company should "model" its financial statements. Such a model causes management to be explicit about a target set of relationships—growth, profits, turnovers, and leverage—toward which the company should work. It also can prove useful in assessing the implications for those relationships of the dynamics of the company's indus-try—exploding or maturing markets, revolutions or evolu-tions in technology, pressures for higher product performance or lower prices, changing distribution channels, and so forth.

Model your firm's financial statements as part of your long-term financial planning.

Table 7-2 outlines alternative financial models for six companies with different characteristics and strategies. These company differences are manifested in the assumptions shown at the top of the table. Analysis of these five dimensions can get the modeler off to a good start on the process of long-range financial planning:

1. Degree of integration (the degree to which the company is self-sufficient [highly integrated] in manufacturing). Highly integrated companies require larger investments in fixed assets and must generate higher gross margins than companies that rely on subcontractors and other suppliers for major inputs to their products.

Analyze these five dimensions to start your long-range financial modeling.

2. Stability of operations. The predictability of the cash flow is a function of the business' susceptibility both to the vicissitudes of economic cycles and to changing competitive conditions. In turn, business stability influences the company's appropriate capital structure.

3. Level of technology. The more the company relies on state-of-the-art technology, the less likely it is to enjoy stable operations and the more it will typically need to spend on development expenses to maintain its technological advantages. Low debt leverage and high gross margins must be sought.

4. Marketing expense. A company that markets components to OEM customers or to distributors must have a lower marketing expense, and therefore can tolerate a lower gross margin, than a company that deploys a direct sales force with substantial application engineering support.

5. Working capital required. The nature of the product, the manufacturing process, and standard terms of sale in its marketplace determine the company's investment in working capital, primarily accounts receivable and inventory.

Each of the hypothetical companies in Table 7-2 has annual sales of $100 (sales can be thought of as an index number) and a return on equity of about 15 percent. Nevertheless, they pursue different strategies, as reflected in variations among their returns on sales (5 percent to 8 percent), asset turnovers (1.2 to 2.2), debt leverage (1.3 to 2.4), and gross margins (25 percent to 56 percent).

Table 7-2. Alternative financial models

	A	B	C	D	E	F
Assumptions						
Degree of integration	High	High	Medium	Medium	Low	Low
Stability of operations	High	Low	High	Low	Low	Low
Level of technology	Low	High	Low	High	High	Low
Marketing expense	Low	Low	Medium	High	High	Low
Working capital required	Medium	Medium	High	Medium	High	Low
Income statement						
Sales	$100	$100	$100	$100	$100	$100
Cost of goods sold	75	59	62	55	44	70
Gross margin	25	41	38	45	56	30
Operating expenses	15	25	26	31	40	20
Before-tax profit	10	16	12	14	16	10
Net income	$ 5	$ 8	$ 6	$ 7	$ 8	$ 5
Balance sheet						
Assets:						
Current assets	$ 40	$ 40	$ 45	$ 40	$ 45	$ 25
Fixed assets	40	40	30	30	20	20
Total assets	$ 80	$ 80	$ 75	$ 70	$ 65	$ 45
Liabilities & owners' equity:						
Current liabilities	$ 20	$ 20	$ 20	$ 20	$ 15	$ 15
Long-term liabilities	27	7	20	5	-0-	-0-
Owners' equity	33	53	35	45	50	30
Total	$ 80	$ 80	$ 75	$ 70	$ 65	$ 45

Company A, a supplier of standardized, mature components, sells to OEM customers and distributors. Being highly integrated, the company has substantial investment in fixed assets and has an asset turnover of only 1.25. Its gross margin is low, but because it needs to spend little on development and marketing, its ROS is 5 percent. Because it is insulated from recessions and other sources of instability, it can profit from high levels of debt leverage (total debt divided by equity = 1.42) and thus achieve a 15 percent ROE.

Company B, unlike company A, makes state-of-the-art products. The resulting low stability demands less debt leverage. Gross margin is now much higher than for company A in order to generate the margin necessary to cover the high development expenses and yield an ROS of 8 percent. The result is again an ROE of 15 percent, despite low leverage (total debt divided by equity = 0.5).

Company C, an instrument manufacturer, operates in a somewhat more mature set of markets and technologies. Marketing expenses are moderately high (direct sales force, but not extensive application engineering) and the mature technology demands only modest engineering expenditures. Once again, stable operations permit high debt leverage (total liabilities are 114 percent of equity). Moderate integration permits lower investments in fixed assets, but the nature of both the product and the manufacturing process demands heavy investment in inventory and accounts receivable. The company is mid range in gross margin, ROS, and asset turnover; higher than average debt leverage permits this company also to achieve a 15 percent ROE.

Company D, a high-technology systems business, is somewhat unstable and requires heavy expenditures for both marketing and engineering development. Gross margins now need to approach 50 percent in order to fund these heavy operating expenses and still permit a competitive ROE while holding debt leverage to a low level.

Company E is a supplier of state-of-the-art systems. Utilizing many purchased subsystems, it has low debt leverage and heavy working capital investments (long work-in-process cycle). It must earn a very handsome gross margin to cover heavy marketing and engineering expenses while earning an adequate ROE. Because of low integration, total asset turnover has now increased to 1.54, in spite of higher investments in working capital.

Company F is a subcontracting assembly shop. Components are purchased from others, and the operations are labor intensive and require little in the way of capital equipment. With in-process manufacturing time short and strong working relationships with customers (rapid payment of accounts receivable), it achieves a total asset turnover of 2.2. Marketing and engineering expenditures are minimal; thus, a low gross margin of 30 percent is still adequate to provide strong ROE, even in the absence of much debt leverage (total liabilities are only half of equity).

Elements of a company's long-range financial strategy must be internally consistent. The firm's financial structure must be compatible with its asset requirements and the stability of its cash flow. Prices (and therefore the gross margin), reflecting the product's technological sophistication, must afford adequate expenditures on marketing and on maintaining technical superiority. The appropriate degree of integration is a function of both stability of operations and level of technology because these two factors influence the firm's capital structure and thus its ability to finance the assets required for integrated operations.

Make the components of your financial strategy internally consistent.

∴ The factors that dictate the structure of both the income statement and the balance sheet are the degree of integration, stability of operation, level of technology, level of marketing expenses, and working capital requirements. Analyze these factors to start constructing a long-term financial plan for your firm. Develop model or target financial statements to help ensure that all elements of the financial strategy are internally consistent.

 # *Evolution of the Financial Function as a High-Technology Company Grows*

The financial function in a high-technology company must change as the firm evolves and matures. The demands on financial managers are particularly subject to change as a high-technology company moves from start-up to a successful,

The chief financial officer's job evolves with the firm.

publicly owned firm, and these changes are related to financing growth, both internally and externally. The chief financial officer must evolve from a cash manager to a competent, internally focused controller. In time, he or she must also assume the role of the externally oriented treasurer.

Burn Rate

Optimize the burn rate in a start-up company.

During the very early stages of a company, particularly during the development phase and prior to initial shipment of products or services, the financial officer should be preoccupied with cash outflow. The monthly outflow of cash in a start-up company is referred to by managers and venture capitalists as the *burn rate*. Optimizing that burn rate—striking the proper balance between accelerating product and market development and husbanding the available cash—is the critical task. Minimizing the burn rate is counterproductive if it delays product introduction for very long.

Start-up companies often exhaust their initial supply of investment capital prematurely. Additional seed capital, or early stage financing, typically proves to be very expensive in terms of ownership dilution to the founders, particularly if development milestones have not been met.

Salaries should comprise the great majority of the monthly burn. Steps should be taken to minimize nonsalary cash outflows, by, for example, renting rather than buying space and equipment, minimizing leasehold improvements, and substituting equity (common stock) incentives for high current salaries.

External Financial Relationships

No company can afford to ignore cash flow as the company grows, but the financial officer of the small, growing firm needs to be increasingly concerned with external financial relationships. Once shipments have commenced, the opportunity for bank financing presents itself, and the financial officer should be concerned with developing a strong and trusting banking relationship. Bankers are more willing to lend to a technology-based company if they understand the company's products and markets, have followed company progress from its early days through initial shipments and customer acceptances, and know, like, and trust the firm's managers. The financial officer should also participate in selecting external auditors and begin to develop a strong working relationship with the audit firm, calling on the theoretical accounting expertise within that firm as needed.

As your firm grows, develop banking relationships and a cost-effective accounting system.

At about this same time, the financial officer needs to create and implement appropriate accounting processing systems. The emphasis here is on "appropriate": neither too elaborate nor too crude. In the early stages, very simple systems may suffice for tracking accounts receivable; but payroll systems, particularly if employees reside in several states, are never simple because complex tax and labor laws exist at both the state and federal levels. Control at this stage of the company's growth can be largely of the "touch and see" variety. Elaborate variance reporting or inventory valuation systems are typically unnecessary. Nevertheless, companies develop habits at this stage. Ideally, these habits reinforce fiscal responsibility (if not frugality) as well as good control of both expenses and assets. Systems should be put in place early on to reinforce these good habits.

Financial Control Systems

As the company continues to grow, controllership activities take on tremendous importance. Delegation of management authority and responsibility becomes more widespread, and the financial systems must adjust accordingly. A higher premium needs to be placed on internal controls and on responsibility accounting. Budgeting now must be decentralized, with an emphasis on "bottom up" budgeting and widespread commitment by middle and top management both to the process of budgeting and to the final budget itself.

Delegation requires control systems based on responsibility accounting.

Asset management now needs to be given high priority because it is at this stage in the growth of high-technology companies when stretched accounts receivable and slow inventory turnover too frequently result in financial crises.

The company is probably now producing multiple products and selling to a wide variety of customers, perhaps in several markets and in several countries. Increased attention should be given to collecting from customers, not just obtaining purchase orders from them, and to inventory reductions, particularly in raw and in-process inventories through the use of MRP systems, closer working relationships with vendors, and similar devices.

Treasurer's Operation

Although the controllership function continues to be critical to the company's long-term succcess, further growth also brings to the fore the role of the treasurer—the externally oriented financial officer. Now relationships with the financial community become broader and more sophisticated, particularly if the company's shares become publicly traded and its business demands international financing. Shareholder relations and public reporting consume substantial time, not only of the treasurer but also of the chief executive. Tax matters assume greater importance as the opportunities to save current tax dollars through sometimes rather elaborate maneuvers can no longer be overlooked. Again, international activities can complicate the company's tax planning.

Further growth focuses attention on the treasurer's job.

As growth continues, financial analysis becomes more sophisticated and comprehensive, and the need for an information data base increases. Although the responsibility for developing and maintaining information systems should not necessarily be lodged with the financial function, it is essential to integrate the accounting system with other elements of the overall information system. The organization's financial managers need to be increasingly responsive to the informational needs, and requests for analytical assistance, that emanate from managers at all levels.

∴ Financial managers of a high-technology firm must be cognizant of the shifts in demands on them as the company evolves from a start-up to a broad-based, publicly owned business. Controllership functions predominate in the early stages, and the treasurer's tasks assume importance as the firm achieves sustained high growth. They must establish external financial relationships, build a financial control system, and integrate the accounting system with other elements of the overall information system.

Highlights

- The appropriate debt leverage for a firm depends on its asset intensity; its stability; market, technology, and other risks; and the risk preferences of managers and shareholders.

- Short-term loans are usually obtained from commercial banks, which typically require collateral (often accounts receivable) and protective covenants.

- Other current liabilities that can serve as important sources of short-term funds are accounts payable and customer down payments.

- Long-term borrowing by high-technology companies that are past the start-up stage typically requires equity kickers.

- Leasing is equivalent to secured intermediate-term borrowing and offers opportunities for expanded low-cost credit.

- The four primary markets for equity capital for a high-technology firm are venture capital, other private placements, the public, and company acquisition.

- A decision to utilize either the public or the acquisition market should be preceded by a careful analysis of pros and cons.

- There is no one right financing strategy for a high-technology firm.

- The degree of integration, stability of operation, level of technology, required marketing expenditures, and working capital requirements dictate the structure of the income statement and the balance sheet.

- To ensure that the elements of your company's financial strategy are internally consistent, develop model or target financial statements as part of your long-term financial plan.

- The financial tasks grow in number and complexity as the company evolves from start-up to sustained high growth.

Notes

1. Stanley E. Pratt, ed., *Guide to Venture Capital Sources*, 6th ed. (Wellesley Hills, Mass.: Capital Publishing, 1982).

Chapter Eight

General Management and Personnel Policies in High-Technology Companies

---

This chapter covers the following topics:

- The General Manager's Job in High-Technology Companies
- The Evolution of General Management as the Company Grows
- Background and Preparation of High-Technology General Managers
- Corporate Culture: Your Firm's Values and Practices
- The Key General Management Challenge: Encouraging Creativity
- Organizing and Rewarding Knowledge Workers: Key Personnel Policy Issues

---

Management is the subject of each chapter in this book. Chapters 2, 3, and 4 focused on functional management—engineering, marketing, and manufacturing. In this chapter, I emphasize general management and those organizational and personnel policy issues that face high-technology companies.

First, what is the general manager's job and are there any peculiarities of high-technology companies that affect that job? Throughout much of this book, I have focused on the boundaries between functional areas of the business. I have emphasized—and do so again in this chapter—the important role general managers must play in orchestrating interfunctional communication and cooperation. But the job is complex and paradoxical.

Second, how does the manager's job change as the small technology-based company grows? We hear of growth plateaus and management crises in high-growth technical companies.

Is there any predictable pattern to these stages and what can management do to prepare for these impending crises? More generally, what background and set of experiences best equip aspiring general managers for success in high-technology companies?

A corporation's culture—a useful but overworked term these days—embodies the firm's management philosophies and principles. Because this set of organizational values serves as the basis for much of the day-to-day decision making at all levels of management, what role can and should general managers play in shaping and changing the corporate culture? Relatedly, are there specific actions or personnel policies that can enhance the firm's creative environment, an essential condition in high-technology companies? Finally, what other specific personnel policies deserve special attention in high-technology companies?

My purpose is not to review the extensive literature on organizational structure and general management; rather, I seek to highlight some dilemmas and techniques that are particularly relevant to those businesses where technology is an important key to strategy.

The General Manager's Job in High-Technology Companies

General management involves a bewildering mix of doing things oneself and doing things through others; of leading, but not getting too far in front; of participating in both ceremonial and more tangible acts; of being at the same time systematic and opportunistic; and of planning and acting. It is, in short, a paradoxical job.

Dr. Simon Ramo, cofounder and top executive of TRW, notes that "A technological company cannot be managed well either by total centralization of decisions and control at corporate level or by the other extreme of total delegation. . . . The essence of good management is a balance, the right hybrid, of these two different approaches."[1] This suggests that the general manager must be willing and able to cope with a fair amount of ambiguity in his or her day-to-day job. One minute the general manager is bold and decisive and later purposely vague and nondirective in order to give middle managers additional elbow room. The general manager must be simultaneously a good delegator and a constructive meddler, a forceful leader and a good listener.

The general manager's job is a set of contrasts.

Managing or Doing

Doing and managing are not the same thing. All individuals who occupy management positions pursue a mix of tasks that involves both doing and managing. When the sales manager is making a presentation to a key customer, he is selling, not managing. When the president writes the quarterly report to shareholders, she is doing, not managing.

During the entrepreneurial phase of the business, the chief executive's tasks are more doing than managing. Failure to alter that mix as the company grows can lead to problems; yet this shift does not come easily for many entrepreneurs and their lieutenants. They like the doing part of their jobs and believe, often correctly, that they are better at these tasks than others in the company. For example, the sales manager is probably correct in believing that he delivers a more effective customer presentation than the salesperson working for him. But, if he makes all the key sales presentations, those working for him will not gain the necessary experience. Moreover, the growth of the business may be stunted because the sales manager is physically capable of making only a limited number of sales presentations in a month.

As your business starts up, doing takes precedence over managing.

As one reaches higher levels of responsibility in the organization and as the organization becomes larger and more complex, one needs to spend proportionately less time doing than managing. Yet Steven Brandt points out that doing (nonmanaging activities) "will almost always usurp the managing work for the following reasons:

Managing supersedes doing as the firm grows.

1. The manager has historically excelled in the specific non-management activity. . . .
2. A nonmanagement activity is often spontaneous . . . the phone rings, the mail arrives, or a subordinate asks for help. . . .

3. A nonmanagement activity usually leads to a clean result. An order is won or lost. A problem solved. . . ."[2]

The high-technology manager must not allow doing to consume all available time at the expense of managing. Keeping a daily time log of activities can help the general manager assess objectively his or her current mix of doing and managing activities. When doing tasks exceed 50 percent of total time for the small company general manager, or 25 percent for the general manager of a larger business unit, the general manager should give serious attention to the need for greater delegation of these tasks.

Leading and Integrating.   The key managing tasks involve leading the organization and integrating the viewpoints and activities of the individual functions of the high-technology company.

General management literature in recent years has placed increasing emphasis on the importance of the general manager's leadership role. The top manager of any operating enterprise sets the style for the company and is responsible for articulating and reinforcing the organization's objectives and values, even if those objectives are set in some collective or consensual manner. The term *corporate culture* has entered the lexicon of management to refer to this set of corporate objectives and values. Corporate culture has proven to be a powerful concept in technically based companies. The general manager's single most important task may be to convey, interpret, explain, demonstrate, and incorporate, in actions taken, that culture.

Leadership involves articulating and conveying the corporate culture.

I have stressed repeatedly the importance of interfunctional cooperation and coordination within a technical company. The ultimate responsibility for integrating the functional (and often parochial) views of engineering, production, marketing, and finance belongs to the general manager. To a greater extent than in most other business enterprises (for example, retailers or financial institutions), high-technology companies demand an orchestration among their key functions, where myopia is likely to lead to conflicting priorities. The orchestra leader is inevitably the general manager.

This integration task is particularly difficult when a functional manager is first promoted to general management. The newly appointed general manager is faced with the important challenge of broadening his or her viewpoints to encompass those of all the functions of the enterprise. This challenge is particularly great in a technical company, where individuals with general management potential are likely to be found in engineering, production, and marketing—all the key functions of the business. By contrast, bank general managers are typically promoted from the bank's lending function, rather than from the trust department or the operations department, because lending dominates the bank's activities. A retailer CEO is almost certain to be a merchandiser, because merchandising is the dominant function in the company. This is not so for technical companies, where there is no single route to top management. The successful high-technology general manager is that individual who, when promoted to general management, can shed any parochialism and effectively integrate across all functions of the technical enterprise.

Only the general manager can integrate functional viewpoints.

Ceremonial Tasks. No general manager can—or should—avoid certain ceremonial tasks. When a general manager pays a courtesy call on a major customer or a major investor, he or she is neither doing—that is, selling or arranging financing—nor managing. The visits are largely ceremonial, even though potentially critical to the long-term health of the enterprise. Serving on the hospital board or raising money for a local charity are other ceremonial duties of the chief executive required by the company's position with the community. Ceremonial duties should not be ignored; they are expected by the general manager's many constituencies: employees, customers, vendors, shareholders, creditors, and the community. However, they should not be permitted to dominate the general manager's time, even though they may be pleasant and ego gratifying. As a general guideline, ceremonial duties probably should consume between 10 and 25 percent of the

Maintain a balance among ceremonial, doing, and managing tasks.

general manager's time, with the higher percentage applicable to the general manager who is also chief executive.

Systematic or Opportunistic

*An effective
general manager
doesn't fit the
stereotype.*

Much management literature preaches that the successful general manager plans, organizes, coordinates, and controls. This successful general manager is pictured as being systematic, reflective, all-knowing, and an issuer of strategic plans and operating policies. Most experienced general managers recognize that their actual practices and life-styles do not seem to fit those descriptions and conclude either that they are not very successful (or at least not model) general managers or that the literature is irrelevant.

Two astute observers of general managers, Henry Mintzberg and Edward Wrapp, have explored these differences between reality and the popular misconceptions regarding managers' activities. I summarize Mintzberg's contrasts in Table 8-1. Wrapp pursued a related theme in a provocative article. He states that the successful general manager does not issue policies or edicts, or even master plans, and does not spell out detailed objectives for the organization. Rather, successful general managers know how to do the following:

1. "Keep open many pipelines of information." They do not rely on either formal information systems or information processed up through the formal chain of command.
2. "Concentrate on a limited number of significant issues." They do not dissipate their energies across the many "time-consuming activities that have infinitesimal impact on corporate strategy."
3. "Identify the corridors of comparative indifference." Wrapp points out that "a good organization will tolerate only so much direction from the top; the good manager therefore is adept at sensing how hard he or she can push."
4. "Give the organization a sense of direction with open-ended objectives." The emphasis here is on effective leadership, giving direction without providing such precise or detailed instructions that the organization's creativity and initiative are stifled.

5. "Spot opportunities and relationships in the stream of operating problems and decisions." The general manager is "a planner and encourages planning by . . . subordinates, but the general manager knows that even if the plan is sound and imaginative, the job has just begun. The long, painful task of implementation will depend on the [general manager's] skill, not that of the planner."[3]

Thus, the general manager has some of the characteristics of both a muddler and an opportunist. Some of this same flavor of the general manager's activities is captured in a phrase that has become popular in high-technology companies in

Be both a muddler and an opportunist.

Table 8-1. Popular misconceptions and facts about managers' activities

Misconception	Fact
The (general) manager is a reflective, systematic planner.	Managers work at an unrelenting pace; their activities are characterized by brevity, variety, and discontinuity; and they are strongly oriented to action and dislike reflective activities.
The effective manager has no regular duties to perform.	In addition to handling exceptions, managers perform regular duties, including ritual and ceremonial, negotiations, and processing of soft information that links the organization with its environment.
The senior manager needs aggregated information, which a formal management information system best provides.	Managers strongly favor the verbal media—namely, telephone calls and meetings.
Management is, or at least is quickly becoming, a science and a profession.	The managers' programs remain locked deep inside their brains. Thus, to describe these programs, we rely on such words as "judgment" and "intuition," seldom stopping to realize that they are merely labels for our ignorance.

Source: Adapted from Henry Mintzberg, "The Manager's Job: Folklore and Fact," *Harvard Business Review* (July–August 1975), pp. 49–61.

recent years: MBWA—management by wandering around.[4] The literature is full of references to MBO, management by objectives—the systematic establishment of objectives at all levels in the organization—but many argue that MBWA is a corollary and valid management process.

Practice MBWA (management by wandering around).

MBWA suggests that effective general managers wander through their organizations—somewhat, although not entirely, randomly and certainly with great purpose. Such wandering permits them to maintain multiple information channels; to obtain and process soft information; to transmit the corporate culture and provide leadership informally, including one-on-one encounters; to rely on the preferred verbal medium of exchange rather than on written and otherwise preprocessed information; to spot opportunities; and to identify what Wrapp calls corridors of comparative indifference.

The act of wandering around is somewhat ceremonial. The general manager's presence, interest, and personal characteristics are demonstrated over time to the entire organization, even though the wandering itself probably appears to be somewhat aimless and highly unceremonial. Where integration across functional boundaries is an important task for the general manager, as it is in technology-based companies, MBWA is a highly effective practice. Even though it appears to be inefficient and time-consuming, it works.

Action Orientation

High-technology general managers should not be afraid to take action. The rapid change that characterizes high-technology environments requires that general managers be action oriented. Some action is generally better than the delay required to study the problem further.

Encourage action rather than further study.

In their recent and widely acclaimed book on excellence in American management, Peters and Waterman say, "There is no more important trait among the excellent companies than an action orientation . . . the evidence of this action orientation seems almost trivial: experiments, ad hoc task forces, small groups, temporary structures. . . . [The excellent companies] are seldom stymied by overcomplexity. They don't give in and create permanent committees or task forces that last for years. They don't indulge in long reports." Peters and Waterman offer some delightful axioms that effectively convey these ideas:

- "Do it, fix it, try it."
- "Ready, fire, aim."
- "Chaotic action is preferable to orderly inaction."
- "Don't just stand there, do something."

They characterize their sample of excellent companies as "experimenting organizations."[5]

Paradoxes

The top management job in high-technology experimenting companies is filled with ambiguity, requiring behavior that is consistent only in its variety. No single, nonexperimenting management style will suffice. Anyone unwilling to contend with the inherent paradoxes should avoid the general manager's job in a technology-based business.

Flexibility and tolerance for ambiguity are essential general manager traits.

Two observers of and participants in high-technology management sum up the need for adaptive changing behavior:

> Managements in the high-technology area must sometimes espouse organizational disorder but at other times, most of the time, espouse order. Disorder, slack, and ambiguity are synonymous with innovation. . . . Slack provides room for the entrepreneurial function. . . . Managers must sometimes disorganize innovation. . . . The successful high-technology firm, then, must be managed ambivalently. A steady commitment to order and organization will produce one-color Model T Fords. Continuous revolution will bar incremental productivity gains. It's knowing when and where to change from one stance to another, and having the power to make the shift, that is the core of the art of technology management.[6]

∴ General managers in high-technology companies are called upon both to do and to manage. Their jobs combine leading, performing ceremonial tasks, and integrating the various functions of the company. Although general managers are popularly viewed as being systematic policymakers, in fact, successful ones often have many of the characteristics of opportunistic muddlers. Good general managers are action oriented, flexible, and able to live with ambiguity.

➡ *The Evolution of General Management as the Company Grows*

In this discussion of the high-technology general manager's job, I have noted several ways in which the demands change or evolve as the technical company grows and matures. General managers themselves either change and grow—acquiring new skills and altering their management styles—or are replaced (or should be replaced, if the directors are alert and responsible). Thus, preparing for the role of general manager is an ongoing process, particularly within rapidly growing technical companies. Fortunately, there seems to be some pattern to these changes, and managers can benefit from anticipating how their management style must change over time.

Observers of smaller technological companies repeatedly cite two truisms:

Unless general managers grow and change, growth is stunted.

- Growth of technical companies frequently plateaus, and restoring growth to such companies is difficult.
- The top executive of a high-technology company frequently has to be changed as the company "outgrows" his or her competency.

These truisms are linked: Outgrowing the CEO's (or management team's) competency causes the plateauing of growth. That is, because their tasks evolve as the company grows and matures, general managers must also grow and change. Failure to do so stunts the growth of the promising technical enterprise.

Plateaus are traditionally linked to company size, measured by sales volumes. Thus, one often hears of the $2 million, the $10 million, or the $100 million plateau. More realistically, but also more subjectively, these plateaus are better defined in terms of organizational complexity. Thus, for example, a producer of a limited line of computer disk drives sold to a few OEMs may be able to generate $75 million of sales with a reasonably uncomplicated and small management organization. Another company, producing a broad line of analytical instruments sold to a diverse set of customers, may find that by the time it reaches $75 million in sales, it has several operating divisions, a number of foreign sales subsidiaries, and a complex financial structure and control system. The disk drive manufacturer may only now require the organizational complexity that the analytical instruments company had at a sales volume of $15 million.

One astute observer of the two truisms just outlined, Larry Greiner, has described the phenomenon in a classic article.[7] Figure 8-1 illustrates the five phases of growth Greiner iden-

Management style must evolve in order to avoid revolutions.

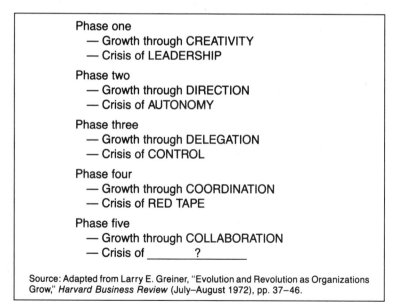

Phase one
— Growth through CREATIVITY
— Crisis of LEADERSHIP

Phase two
— Growth through DIRECTION
— Crisis of AUTONOMY

Phase three
— Growth through DELEGATION
— Crisis of CONTROL

Phase four
— Growth through COORDINATION
— Crisis of RED TAPE

Phase five
— Growth through COLLABORATION
— Crisis of _____?_____

Source: Adapted from Larry E. Greiner, "Evolution and Revolution as Organizations Grow," *Harvard Business Review* (July–August 1972), pp. 37–46.

Figure 8-1. Evolutionary stages and revolutionary transitions

tifies and his labels for each of the evolutionary stages and for the revolution, or crisis, that frequently attends the transition from one stage to the next.

Creativity and the Leadership Crisis

Greiner identifies the first stage as one of creativity, a time when little formal management is required because the operating team is small, highly motivated, entrepreneurial, very hardworking, and in extremely close contact with one another. Both output and personal satisfaction are high. The several-person technical team that is developing (and perhaps producing in small quantities) a distinct new product for a niche market—and doing so within a start-up company—is a prevalent example of this stage. Small partnerships of professionals—consultants, lawyers, or doctors, for example—also fit into this stage.

A strong leader must emerge during the creativity phase.

The crisis that befalls these first-stage companies as they grow is that of leadership. Demands on the company become broader and include manufacturing, financing, customer service, and selling. The highly motivated original team is now supplemented with personnel having less intense dedication and for whom the information communication systems are inadequate or inappropriate. Confusion ensues. To eliminate the confusion and set a direction for the company, a leader must emerge. Fortunately, this leader is often present within the founding team. Not so fortunately, other founders may resent the greater authority thrust upon or assumed by the leader and may leave the company, grumbling that "working here is not as much fun as it used to be."

A number of venture capitalists have explicitly recognized the importance of this transition from stage one to stage two in the way they have organized and financed some high-technology start-ups. They have financed a founding team of technical experts to develop the product through the prototype stage—Greiner's creativity stage. At the end of this stage, assuming success, the venture capitalists arrange to provide two important ingredients to the embryo company: more money to finance manufacture and sale of the product; and leadership in the form of a CEO brought into the company from the outside.

Direction and the Autonomy Crisis

During Greiner's second stage, growth through direction, sustained growth results from strong and directive leadership. A somewhat more formal management structure is now in place, based upon a classical functional organization. Communication formalizes, but the CEO makes virtually all the key strategic and policy decisions.

As the business grows, the CEO becomes stretched and is unable to direct personally the diverse activities of the business. Lower-level managers find their restricted decision-making authority frustrating and cumbersome. The result is either (or both) a leveling of growth or a crisis of autonomy, as the lower-level managers demand more authority or leave the company. The solution is typically greater delegation of authority, a solution that is often incompatible with the highly directive and domineering management style of the executive who led the company in phase two. Not surprisingly, the board of directors frequently institutes a change in CEO as it attempts to deal with the problem of slowed growth.

The growth through direction stage is often followed by an autonomy crisis.

Delegation and the Control Crisis

It is at the control crisis point that the entrepreneur may depart the scene, turning over the management reins—whether eagerly or reluctantly—to a new CEO (or team) whom the board entrusts to bring "professional management" to the company. Decentralization occurs, with more authority delegated downward in the organization: Top management concentrates on management by exception and on dealing with the external environment. Such decentralized organizations place high demands on formal information and control systems, and failure to develop these systems apace with company growth may bring on the third crisis: the crisis of control.

The struggle now is the classic one between centralized and decentralized management. Lower-level managers have had a taste of autonomy; the top management team, having been "burned" by out-of-control conditions in remote corners of the now far-flung organization, is anxious to reassert more direct control over the total company. As this struggle pro-

The growth through delegation stage brings on a struggle between centralized and decentralized management.

gresses, another growth plateau is likely to ensue. Companies that continue to progress from this point adopt techniques to effect coordination without heavy-handed control.

Coordination and the Red Tape Crisis

During phase four, coordination, divisions are gathered into "groups"; corporate staff activities typically expand; formal planning procedures are instituted; and cross-functional and cross-divisional committees abound. If or when the proliferation of systems, programs, and committees "begins to exceed its utility, a red-tape crisis is created. Tension builds between line and staff organizations, and between headquarters and field operations. Procedures take precedence over problem solving, and innovation is dampened. In short, the organization has become too large and complex to be managed through formal programs and rigid systems."[8]

Ineffective coordination can result in excessive red tape.

Thus, excessive red tape may be another cause of a growth plateau. Greiner sees the way out of this crisis into the final phase, which he labels collaboration, as emphasizing "greater spontaneity in management action through teams and the skillful confrontation of interpersonal differences. Social control and self-discipline take over from formal control."[9] Characteristics of this phase are a focus on solving problems through team action, greater emphasis on matrix organization structures, willingness to experiment with new practices and procedures, and increased prominence of educational programs. Greiner only speculates on the crisis that may ultimately befall the firm that has successfully implemented a collaborative organization.

Most participants in, or observers of, high-growth technical companies struggling with the various problems of growth subscribe to a "phase and crisis" model such as Greiner's. The names may be different, but the observation that demands on management change dramatically as the company grows and matures seems irrefutable. To be forewarned is to be forearmed. Managers and boards of directors of high-technology companies should recognize these growth stages and effect the changes that are necessary to maintain company growth and organizational health, even when these changes involve the unpleasant task of replacing or reassigning individual managers, including the president.

∴ As small companies grow, the demands on general managers evolve from an emphasis on personal creativity to a series of stages in which managers must emphasize firm direction and then delegation. As the firm becomes large, coordination and collaboration characterize the management style. By recognizing that the transitions between these stages are frequently preceded by a growth plateau and accompanied by organizational turmoil, top managers can take early action to install the organizational structure and instill the management style appropriate for the next stage.

Background and Preparation of High-Technology General Managers

A technology-based business is one whose strategy is in major part keyed to technology; so its general managers—that is, the company's key strategists—must be conversant with the technologies to which the company is or might be linked. Therefore, most successful general managers have a strong technical background, whether achieved by formal education, experience, or self-study. Because the rate of technology change may shorten the useful life of technical facts learned in a university to ten years or less, self-study of technological developments is a key part of general managers' jobs.

A strong technical background, augmented with continued study, is essential.

Most top executives in U.S.-based, multinational, multiindustry companies, operating predominantly in mature industries, are not very technologically literate. Their orientation is primarily financial. Such preoccupation with finance is generally inappropriate for the high-technology company. The key and constraining resources in a high-technology company are not financial; rather, they are technology and people. General managers, therefore, find that the key skills they need are knowledge of the technology and the ability to lead, motivate, and supervise people.

The high-technology general manager can ill afford to be a narrow technologist. Although technology is *a* key to the strategy of the high-technology business, it is seldom the sole key. To rephrase an old cliche: Seldom will the world beat a path to your door simply because you have a better mousetrap. Even if customers beat a path to observe the startling new product design, the product still must be engineered, produced, sold, and serviced, and the enterprise must be financed. A dishearteningly large number of entrepreneurs with superb technical credentials, but little or no management expertise, have been frustrated in their attempts to build companies to capitalize on their technological breakthroughs.

But knowledge of the market and the art of people management is also necessary.

The successful general manager of a high-technology company, then, needs to be well grounded in the market, the technology, and the art of people management. The key strategic decisions facing most technical companies are at the intersection of technology and the marketplace—that is, as stressed in Chapter 2, they involve the selection of those product-market segments that the company should pursue. Nevertheless, the best technical and market strategies still require implementation, and implementation (as I have emphasized throughout this book) requires the artful orchestration of the efforts of personnel across all functional areas of the business.

Preparation

Experience early in one's career in the following two critical activities is a useful start in preparing for this demanding position in a high-technology company:

1. Selling—meeting the customer, getting to understand his or her needs, and relating the selling message to those needs
2. Supervising people—leading an engineering development team, a production line, or a sales region

Too frequently, highly educated future general managers—MBA graduates or Ph.D. research engineers—never gain these basic experiences. They spend their early years with the technical company in staff management positions (often planning or financial analysis) or deep in the laboratory. Their personal interactions are limited to other professionals and seldom involve hourly workers, customers, or engineering personnel. The absence of this nitty-gritty experience can result in a severe experience gap when these otherwise well trained individuals later move into general management positions.

Learn early in your career how to sell and supervise.

These comments suggest the desirability of several changes in job assignments, or what the personnel professionals call *job rotation*. Job rotation is important in preparing for any general management position, but it is doubly important in high-technology companies because interfunctional coordination and communication are critical. Some outstandingly successful companies, such as Hewlett-Packard and Raychem, have made a practice of regularly moving promising managers across functional boundaries. For example, the current vice-president of research and development at Hewlett-Packard has in recent years had assignments as division general manager, director of corporate planning, and head of the personnel department.

Job rotation is especially important in high-technology firms.

Experience as a general manager of an operation of limited scope is good training for future general management positions of greater scope. To this end, product management positions can offer a quasi–general management experience for young professionals. Typically, a product (or market) manager is responsible for coordinating all the company's activities with respect to a particular product line (or market). The product manager has limited authority, almost no direct supervisory responsibility, but a kind of quasi–profit responsibility. A product manager's key function is to coordinate across functional lines—between engineering and marketing, between engineering and production, and between marketing and production. The product manager must attempt to optimize among conflicting demands and priorities within the organization while keeping careful watch over customers' needs and satisfaction—a reasonable definition of the general manager's job.

Gain Key Experiences

In preparing for general management responsibilities, the aspirant and his or her mentors should focus on the key ingredients for success within the particular high-technology business and be certain that experience is gained in those key areas. This attention to key areas of experience should, of course, also preoccupy top management and the board of directors as they select future CEOs and general managers.

Gain experience in areas key to your company's future success.

If efficient, high-quality production is key to the company's competitive position, experience in manufacturing is essential. If international sales are key, and the company finds it necessary to adapt its products for the different cultural traits of the countries where it operates, some international experience—perhaps including having lived in a foreign country—may be desirable. If a strong distributor network is the key element of the company's marketing mix, the aspiring general manager had better be certain that he or she has had one or more positions that involved exposure to these distributors, their people, their motivations, and their operating procedures. If the company's key customer is the government, the future general manager would be foolish to restrict his or her experience to the commercial market.

It is distressing how frequently controllers who have never called on customers or engineering managers who have not ventured far from the laboratory (and are even contemptuous of manufacturing personnel) feel that they have prepared adequately for general management positions. Simply doing one's own narrow job well and observing one's counterparts in other functional areas of the business is seldom adequate preparation for general management.

∴ Top managers in high-technology companies must be conversant with and keep current on their companies' technologies. Most aspiring general managers are engineers and should have held positions in those company functions that are key to its success. Because of the importance of integration across functional boundaries, high-technology companies should see to it that prospective top managers gain experience in several functions.

Corporate Culture: Your Firm's Values and Practices

The culture that develops within a corporation both *limits* the general manager's role and influence and presents an opportunity to *extend* his or her effectiveness. Corporate culture is the set of values held by the organization and the set of practices that derive from them. These values derive from assumptions about, for example:

1. Employees, their motivations and their personal and professional expectations
2. Customers, and how they are to be served
3. Suppliers, and how they are to be treated
4. The community, and the company's relationships with it
5. The degree of risk the company should assume and its effects on relations with employees, financial markets, suppliers, and customers
6. The relative importance of the enterprise's various functions (for example, whether the business is marketing driven, technology driven, or financially driven)

Culture builds from assumptions about the corporation's employees, customers, suppliers, shareholders, and community.

Corporate culture both shapes and is shaped by business strategy. A strategy that relies on market share and scale economics for success inevitably implies a different corporate culture than one that depends on the rapid creation and introduction of new products to fill niche markets, in turn created by the company's imaginative marketing department.

Corporate cultures are real; they can be sensed within the first few minutes that one is in the facilities of a company with a strong and well-understood culture. A strong corporate culture guides employee decisions, reducing the need for multiple approvals and excessive checking and reporting. Hewlett-Packard has a strong and operative corporate culture that is understood throughout the company; its implementation is referred to as the "HP way" of doing things.

Top managers define and communicate corporate culture by their words and deeds.

Managers, most particularly senior managers, define and transmit corporate culture. Consciously or unwittingly, top managers define the corporation's culture by their pronouncements and, even more, their actions. The demonstration and transmission of corporate culture is a major task of top management in well-managed companies.

A clear, concise, and understandable written statement of corporate objectives helps define and communicate the corporate culture. Such statements, however, need to be free of both jargon and excessive overlay of financial and other numbers. The statements instead must articulate simple guidelines as to, for example, the role and worth of the individual, the attitude to be taken toward customers, and the relative priorities that will be attached to growth, returns to shareholders, employment stability, and financial risk.

The influence of corporate culture is more likely to be strong and useful in a company where the same culture is appropriate for all divisions or groups. Thus, a multiindustry company (a conglomerate) is unlikely to develop a single corporate culture. When activities range from high-technology manufacturing to financial services to retailing, as they do at such companies as Teledyne, ITT, and Litton, divisional or subsidiary cultures may develop, but they are not likely to be strong and there will be little similarity among these subcultures. This condition represents a real handicap for conglomerates: Their top managers have difficulty articulating effectively— and acting consistently with—a broad set of partially conflicting cultures.

Building Cultures

Because a well-understood and accepted corporate culture can provide powerful decision-making leverage to the organization, alert managers of high-technology companies should strive to build a strong corporate culture. But they must realize that it takes time. A culture is not created by edict. It evolves over a period of years or decades and is strongly influenced by both the words and deeds of the chief executive and other members of top management.

Building a corporate culture takes time.

Managers must also realize that both words *and* deeds count. When the deeds do not match the words, corporate hypocrisy, not corporate culture, results. When, for example, the worth of the individual and assurances of stable employment are proclaimed during periods of prosperity and layoffs occur during periods of recession, management's true attitude toward employees is transmitted by deed rather than by proclamation. When the company's suppliers are assured by the stated corporate objectives that they are viewed as key, trusted partners with the company, but then are treated rudely by purchasing and paid erratically by accounting, both suppliers and company personnel quickly learn that the stated corporate objectives are a mockery. Strong corporate cultures are built by consistent actions—that is, actions that are consistent over time and consistent with the stated company objectives, precepts, and policies.

Avoid corporate hypocrisy.

Changing Cultures

Strong corporate cultures have a great deal of momentum and are thus not easily changed. In general, this condition is beneficial. Most corporate cultures can and do tolerate substantial shifts in corporate strategy. But when a company must radically alter its strategy—in response to abrupt and decisive changes in economic, political, and competitive climates—some change in corporate culture may also be in order. The prevailing corporate culture, the one derived from and supportive of the old strategy, may be somewhat inappropriate for the new strategy. (Note, for example, that Greiner's evolutions and revolutions as organizations grow imply some evolution in the corporate culture [see Figure 8-1].)

A strong culture changes slowly—a desirable characteristic, except when radical changes in strategy are called for. Consider, for example, the telephone companies of this country as they move from a highly regulated to an unregulated environment. The old culture of customer service without undue concern for cost and with no attention to competition is at best suboptimal in the telephone companies' new, highly competitive environment. Customer service is hardly irrelevant, but greater attention must be paid to cost-benefit analyses from the customers' viewpoint, and the function of marketing must assume a more prominent role.

Corporate cultures change slowly.

IBM, too, has had to change its strategy as new competitors have attacked the company, both in very large mainframe computers and in personal computers. From a time when price was seldom, if ever, used as a competitive weapon, IBM has moved to a strategy that admits price—and the threat of drastic price changes, in hardware, software, and service—as a key element in the marketing mix. This change in strategy away from a preoccupation with stimulating primary demand has required that personnel throughout the organization assume a more "scrappy" attitude toward affecting selective demand.

Modifying cultures requires extra management effort.

When top management initiates sharp changes in strategy, and in the accompanying corporate objectives, policies, and practices, it must tend also to the need to change old behavior patterns throughout the organization—behavior that is consistent with the long-prevailing but now less-than-optimum culture. Because the old culture is slow to change, top managers should devote extra effort to articulating the necessary changes in objectives and specific changes in behavior. The message needs to be delivered and demonstrated repeatedly.

∴ A corporate culture is that set of beliefs, values, and policies that both shapes and is shaped by a company's business strategy. Consistent and broadly understood cultures are effective decison-making aids throughout the organization. Managers must understand that cultures take time to build and are hard to change.

 # The Key General Management Challenge: Encouraging Creativity

High-technology companies must seek a corporate culture that encourages and prizes creativity. Indeed, a major challenge for top managers of high-technology companies is to assure the continued creativity of the organization and all its people. Because a high-technology company must, almost by definition, create new products and new markets, it must also attract, stimulate, motivate, and retain individuals who are creative, adaptive, and willing to take personal and intellectual risks. In short, high-technology companies are greatly dependent upon "knowledge" workers.

Stimulate a creative climate for knowledge workers.

Knowledge Workers

Peter Drucker, a prolific writer on management, says:

> A primary task of management in the developed countries in the decades ahead will be to make knowledge productive. The manual worker is yesterday. . . . The basic capital resource, the fundamental investment, but also the cost center for a developed economy, is the knowledge worker who puts to work what he has learned in systematic education, that is, concepts, ideas, and theories, rather than the man who puts to work manual skill or muscle. . . . We still cannot answer what productivity is with respect to the . . . knowledge worker.
>
> Managing knowledge work and knowledge workers requires exceptional imagination, exceptional courage, and leadership of a high order. . . . The knowledge worker, except on the very lowest levels of knowledge work, is not productive under the spur of fear; only self-motivation and self-direction can make him productive. He has to be achieving in order to produce at all. . . . Achievement for the knowledge worker is [hard] to define. No one but the knowledge worker himself can come to grips with the question of what in work, job performance, social status, and pride constitutes the personal satisfaction that makes a knowledge worker feel that he contributes, that he performs, that he serves his values, and that he fulfills himself.[10]

Knowledge workers' range of capabilities is substantially greater than that of "muscle" workers. The best and most creative of the knowledge workers are many times more productive for the company than are average workers.

Knowledge workers are much in demand and highly demanding.

These exceptional knowledge workers are scarce, a condition that, for several reasons, is likely to continue. A substantial shortfall in the awarding of engineering degrees is projected over the coming decade or two. At the same time, the computer revolution will assure a continued shortage of digital engineers, software designers, and, most critically, engineers conversant across the spectrum of technologies involved in integrated electronic systems.

This shortage, in turn, reinforces the current tendency of these highly creative individuals to be independent, even arrogant. They frequently become almost free-lance professionals, demonstrating loyalty to their profession, themselves, and their particular technology, but very little to their employers. Some engineers in Silicon Valley in California think nothing of changing companies, particularly because they can sometimes do so without even changing parking lots. As long as these key technologists are offering their talents in a sellers' market, they will seek that environment where they can be most productive, where they will not only use their knowledge but add to it.

You need creative teams as well as individual stars.

To compound the challenge, the successful technology-based companies require not simply a few super scientists but a whole stable of creative individuals and teams, and not solely in the engineering laboratory but also within the marketing function and on the production floor. Most truly significant innovations do not represent revolutions in technology carried out by a single individual. Rather, they are the product of many small innovations—the evolution of technology—brought forth by many individuals. Important new products, then, are arrived at step by step over time, by teams of creative knowledge workers.

Creative individuals seek the challenges that accompany rapid growth.

These teams seek growth and excitement. Most technical companies find that staff creativity represents the greatest pressure for growth and, at the same time, the most severe limitation to growth. That is, growth in a rapidly changing technical field demands that the ranks of creative engineers, marketeers, and others be expanded without sacrifice in quality, that is, creativity. (Hiring mediocre personnel when the pressure is on will surely drive out real performers.) These same creative individuals demand growth, for growth brings new challenges—both management and technical challenges—and provides the necessary opportunities for expanding their knowledge base. Growth requires the hiring of new personnel, and these new employees bring stimulation and new ideas to the company and its technical staff. Thus, for many high-technology companies, the ultimate "governor" on growth—setting both upper and lower limits to the growth rate—is the rate at which additional technical and managerial personnel can be identified, hired, and assimilated into the organization.

Enhancing the Creative Environment

A creative environment—or, more precisely, a work environment that encourages and stimulates creativity—is not brought about by decree. The experience of many companies suggests at least six helpful actions or attitudes: stimulating the right attitude, balancing objectives, tolerating "extracurricular" work, fostering risk taking, valuing "champions," and balancing the staff.

Stimulate a New Product Attitude. Management can foster, as part of the corporate culture, an emphasis on new products, the expectation that new products will be brought forward to the market in a steady stream. Rather than a new product being an exceptional event, the absence of new products is the exception. Both 3M (Minnesota Mining and Manufacturing) and HP have fostered this attitude, and both have an enviable record of developing and introducing new products. Many of these new products are not revolutionary, and some, particularly at 3M, may even have appeared trivial at first. However, the companies find that sales of some of these minor new products (or product improvements) have greatly exceeded initial expectations.

Create expectations for a steady flow of new products.

Balance Development Objectives. Keeping too short a leash on development engineers—that is, demanding that all work have a potential near-term payoff in the marketplace—is likely to truncate important technical inquiry that could have significant long-term benefits. However, the engineering department operates in an environment where technology is moving fast, product life cycles are short, and competitive threats are serious. Pressure must be maintained for the rapid development of "fixes" to overcome shortcomings of existing products and for speedy movement of new innovations from the laboratory to the market. Creativity is likely to be enhanced by a balancing of the short- and long-term priorities, as each set elicits different types of developments.

Balance short- and long-term priorities.

Tolerate "Bootlegging." A technical organization with some built-in slack is more creative than one where every move is programmed and scheduled. This fact does not imply that technical efforts should not be planned or that schedules should be ignored, but rather that plans and schedules should consciously allow for some slack time. Creative organizations tend to spend slack time in pursuing exciting new technological options that have not yet been sufficiently defined so that they can be submitted to management for formal funding and scheduling. Slack time is spent on so-called "bootleg" projects or activities. (Even when slack time is not built into the system, bootlegging is likely to occur, to the detriment of established schedules.) Where communication between marketing and engineering, and among various engineering sections, is both extensive and constructive, such bootlegging is very likely to be well related to the company's mainstream activities; the understanding of needs stimulates the creative juices. Some slack in the marketing department is also likely to lead to useful bootleg investigations of new market niches.

Build in slack time.

Texas Instruments has attempted to bring bootlegging out of the closet. Anyone with an embryo idea can seek modest funding (typically up to $25,000) from any of a number of sources within TI without elaborate justification of the project. This modest sum permits the individual to take the next step in proving (or disproving) the idea and formulating a proposal for sustained funding.

Foster Risk Taking. Encouraging risk taking is easier said than done. Rewards must be more prevalent for those who take risks than for those who avoid them. At the same time, failure can hardly be encouraged, although probabilities dictate that some failures will be the inevitable result of taking risks. Management can take steps to assure that excess risk aversion does not permeate the organization or that risk adjustments are not compounded as a proposed project undergoes multiple reviews. Young engineers can be evaluated on actions and risk-taking attitudes, and not solely on achievement of successful results.

Reward risk taking, not just success.

Management can inquire about those projects or avenues of investigation that were *not* taken—that is, projects that were turned down because of either lack of promise or excessive risk. When managers must pay as much attention to justifying the rejection of promising ideas as to accepting projects, risk taking is likely to be encouraged. A calculated risk that is taken but proves unsuccessful must not lead to punishment for the risk taker.

The risk that a failure will not be acknowledged promptly is as serious as the risk that a promising but problematic opportunity will not be seized. An innovation that is, at best, mediocre consumes resources that can be better deployed elsewhere. But where risk taking is not highly valued and failure carries a stigma, the innovative engineer, product manager, or division head is slow to accept and act upon the reality of a failed development project, process innovation, or test market. Owning up to "near successes" that will never be real successes must be particularly encouraged.

Encourage early cancellation of near successes.

Value "Champions." Many a new product or new project idea would never see the light of day were it not for the zeal of its champion, that individual within the organization willing to take the psychic and career risks associated with promoting the new idea. The more radical the idea, or the more it flies in the face of conventional company wisdoms ("we tried that once," "we don't do things that way here"), the more zealous—and even obnoxious—the champion must be.

Senior managers have the clout to be champions, but they seldom have the technical and market involvement that results in extreme dedication to the new idea. There are exceptions. Dr. Land at Polaroid championed the instant movie camera (unfortunately, as it turned out), and Mr. Hewlett championed the handheld calculator at HP. Any good entrepreneur is by nature a champion.

But champions are needed among the engineers, product managers, and other creative individuals well below the senior management ranks. When these champions emerge, they should be listened to and encouraged. They should be accorded several avenues of access to key decision makers in the company to ensure against their ideas being judged prematurely by a supervisor who may be both insecure and risk averse. A frustrated champion will soon lose all creative spark or, more likely, leave the company. Most founders of new high-technology companies were once frustrated champions in larger companies.

Encourage personnel at all levels to be champions.

Balance the Technical Organization. Brilliant and creative scientists or engineers are necessary members of the technical team. But if creativity is to be fleshed out into innovations that have market and profit consequences, a mix of talents must be included on the technical staff. Ed Roberts at M.I.T. identifies five staff roles that must be filled if "innovative ideas are to be generated, developed, enhanced, commercialized, and moved forward in the organization":

Include scientists, entrepreneurs, managers, sponsors, and gatekeepers on your technical staff.

1. The creative scientist or engineer, "about whom so much—perhaps too much—has been written"
2. The entrepreneur, often the champion, who pushes forward the idea toward commercialization
3. The project manager, who focuses on specifics and directs and coordinates the effort
4. The sponsor, the senior individual who provides coaching and support (As Roberts says, this role is "that of protector and advocate—and sometimes bootlegger of funds.")
5. The gatekeeper, who brings essential information (technical or market information or both) into the organization[11]

Managers interested in promoting commercial creativity, as contrasted with technical curiosities, should focus on building a balanced team of engineers and managers.

∴ Top managers in high-technology companies must strive to maintain and enhance creativity, particularly among the knowledge workers in their organizations. They must stimulate the acceptance of change, balance short- and long-term objectives, tolerate some "bootlegging" of promising new ideas, foster and reward risk taking, encourage champions of new ideas, and build balanced technical staffs.

Organizing and Rewarding Knowledge Workers: Key Personnel Policy Issues

Certain key decisions general managers make to stimulate creativity, define and reinforce the corporate culture, and encourage interfunctional communication and cooperation become expressed in both the organizational structure and the dominant personnel policies. I focus here on key personnel policies that have been hotly debated in recent years. These policies are key by virtue of the following characteristics of technology-based companies:

- High percentage of professional employees—knowledge workers—as compared with hourly workers
- Rapid change, preferably in the direction of growth, but sometimes resulting in rapid decline
- Shortage of skilled personnel and widespread personnel pirating by the many entrepreneurial ventures that populate the high-technology industries
- Complex organizations and problems that require professionals to interact effectively across functional boundaries

Matrix Management

The classic form of hierarchical, functional organization, although probably necessary, is almost surely not sufficient in a high-technology company. Virtually all companies have some elements of matrix management. That is, some persons in the organization have two or more bosses. In high-technology companies, matrix management is more the rule than the exception, regardless of whether the reporting relationships are formally labeled "matrix."

Matrix organizations are both useful and inevitable.

The most straightforward example of a matrix organization is an engineering department that organizes both by technical specialty and by project. Each design engineer is typically a member of the technical group that comprises his or her particular expertise and of the project team to which he or she is currently assigned. The engineer has two bosses: a group boss and a team boss. The project team leader also has two bosses: a senior manager in engineering and a product (or product line) manager elsewhere in the organization. Consulting and engineering design firms are almost all organized in such a matrix fashion.

The sales engineer, who is both a product specialist and a member of a regional sales office, also operates in a matrix environment. He or she is responsible both to the product sales manager (perhaps located at the home office) and to the regional sales manager. The sales engineer's performance must inevitably be appraised by both "bosses." Similarly, the manufacturing engineer has multiple reporting relationships—to the head of the manufacturing engineering department, to the design engineering project leader, and to the manufacturing manager whose production problems the manufacturing engineer is attempting to fix.

Invest the time to make matrix management work.

Formal matrix management recognizes, highlights, and even demands that cross-functional interaction be carried out. It facilitates communication, permits trade-offs to be made lower in the organization, and provides individual employees with broader exposure to and understanding of other functions and other managers. However, the potential for inefficiency, excessive numbers of meetings, and individual employee frustration is high. Thus, management must be willing to invest the time to make it work and to adopt a conducive management style.

Authoritative management styles are bound to be disruptive in a matrix organization. An employee who reports simultaneously to a tyrannical boss and to one whose style is more participative has a difficult time making the trade-offs in conflicting priorities that pervade such an organization. For obvious reasons, trade-offs will be shaded in the direction of the tyrannical boss and will result in less-than-optimal decisions for the company as a whole.

Thus, in a matrix organization, middle managers must stress listening, relating, coaching, and coordinating and downplay direction and traditional supervision. Performance appraisal becomes the mutual responsibility of two or more bosses. Authority and responsibility inevitably become somewhat diffused. Individual employees must be able to handle a fair degree of ambiguity; management must be dedicated to reducing that ambiguity to acceptable levels and must also be alert to situations where the frustration level for the individual employee becomes debilitating.

Matrix organizations require managers to listen and coach more than direct and control.

This style and practice of management is not easy for some people, particularly highly directive and demanding entrepreneurs. But all organizations—and particularly those in high-technology industries—inevitably contain elements of matrix management. The resulting demands on managers thus cannot be escaped, even in the most strictly functional and authoritative organizations.

Dual Ladders

Virtually all companies whose strategy is based on technology quickly perceive the need to recognize and reward not only key managers but also key technical contributors (scientists or engineers) who carry no (or minimal) management responsibility. The result is the widespread use of so-called *dual ladders:* separate but parallel titles and compensation steps for managers and for technical personnel.

Use parallel compensation and recognition plans for technical and management staffs.

Equity in pay is not enough; dual ladders need to involve both compensation and recognition. Recognition is the more difficult reward to implement. By their nature, top managers are visible. To them flows much of the ego gratification associated with company success: Their names appear in the newspapers; their pictures are prominent in the annual report; and they become known and recognized both within and outside the company. At the same time, for competitive reasons, the contributions of key technical personnel may have to be kept quiet for a considerable period after a technical breakthrough occurs.

Not everyone responds to the same types of rewards.

With some conscious attention, however, the company can see to it that top scientists and engineers are widely recognized within the company for their contributions. Titles, such as senior scientist, can help. Publicity regarding patents granted is another avenue. But, of course, highly creative technical personnel may respond to different reward structures than top managers. To them, the opportunity for more slack in their project assignments so as to pursue particularly intriguing technical developments may be infinitely more important than the ego gratification that attends internal or external publicity about their accomplishments. Fortunately, such slack time may also be very much in the best interests of the company, as discussed earlier. IBM has been particularly keen on this type of recognition for its top scientists and engineers.

In spite of the obvious need for multiple (at least dual) ladders for promotion, compensation, and recognition, the record of success at actually implementing dual ladders in U.S. technical companies is fairly dismal. Scientists themselves do not have the managerial authority to cause this duality. Perhaps, given the culture of the United States and the fact that the impetus for dual ladders must come from top management, it is not surprising that the great rewards—both monetary and psychic—continue to be highly skewed toward senior managers.

Stable Employment

The costs of fluctuating employment levels are high.

Change is the theme of high-technology companies. This change frequently results in pressures for rapid changes in employment levels—massive hiring and massive layoffs. The financial, human, and organizational costs associated with widely fluctuating employment are enormous.

What should be the company's policy with regard to guaranteeing its employees stable employment? In return, what should the company expect from its employees in terms of loyalty? These two questions must be linked. Amazingly, many managers of high-technology firms complain bitterly of the absence of employee loyalty. They indict employees who move across the street (or even across the country) for a 10 percent wage increase or for a slightly more challenging position. Such moves are cited as evidence of disloyalty, or loyalty solely to one's self or to one's profession, rather than to one's employer. Yet these same managers assume no responsibility for stable employment for these valued employees. They see no alternative but to cut employment levels in time of economic stress or competitive pressure. They cite their obligations (loyalties?) to shareholders, which may really translate into concern for their own management positions.

Don't expect employees to be loyal to a company that isn't loyal to them.

The answer is not easy. Lifetime employment in Japan is often cited as the ideal. In spasms of enthusiasm, particularly in strong economic times, managers of American high-technology companies frequently pronounce that stable employment has been adopted as a key corporate objective. But the United States is not Japan; we do not have the bank support, government-industry relationships, subcontractor network (where stable employment is anything but guaranteed), or cadres of "temporary" employees that characterize Japanese industry. In fact, in Japan, lifetime employment (only to age fifty-five) is assured for a very small percentage (estimates are 10 percent) of the total labor force, including almost no women.

A company whose strategy focuses on market share and that builds capacity in anticipation of uncertain demand should face facts: It has great difficulty guaranteeing lifetime employment to its staff. A company that, as a matter of conscious policy, finances rapid growth with high debt leverage also has a difficult time maintaining employment levels when sales turn down. When employment is guaranteed, labor becomes a fixed, rather than a variable, cost, thus increasing operating leverage. High operating leverage and high debt (financial) leverage are a risky combination in a volatile environment.

Reduced turnover improves efficiency and aids recruiting.

High-technology companies have much to gain from stressing stable employment. An immediate and tangible return is reduced company contributions to the state's unemployment insurance fund. Longer-term payoffs, some of which are difficult to quantify, are substantially greater.

1. Stable employment reduces training costs, and training—both formal and informal or on the job—costs enormously, particularly in technology-based companies, where skill levels are high.
2. Stable employment builds company loyalty, with the result that voluntary employee turnover is substantially reduced.
3. Reduced turnover decreases the load on the personnel department, the training department, and middle managers who carry the responsibility for interviewing, hiring, and on-the-job training.
4. Stable work groups—ones where members are not constantly coming and going—are inevitably more efficient.
5. Stable employment reduces the potential for an adversarial relationship to build between management and production workers, the kind of adversarial relationship that has led to strong and inflexible labor unions in traditional U.S. industries.
6. Because new employees seek stable employment, recruiting to fill new positions created by growth becomes both easier and more successful: Top prospects are attracted to the company.
7. Average wage rates can be lower at a company that delivers on its promise of stable employment, although few employers will admit to capitalizing on this advantage.

Consider carefully before publicly committing your firm to a stable employment policy.

Delivering on the promise of stable employment is key. Pronouncing a policy of avoiding layoffs and then being forced into a layoff at the next economic downturn represents corporate hypocrisy that has repercussions well beyond the policy itself. Thus, high-technology companies should think long and hard—analyzing carefully all elements of their corporate strategy, competitive position, management style, and overt or implied promises to the financial community—before committing to a policy of stable employment.

Unquestionably, the wholesale hiring and laying off of personnel is becoming progressively less acceptable in this country. Almost no other highly industrialized country permits—legally or culturally—the wide fluctuations in employment that characterize American industry. Already within U.S. high-technology centers—the West, the Sun Belt, and parts of New England—the practice occurs less than it does in the midwestern home of the mature, highly unionized industries of the country, particularly auto and steel. It behooves managers of high-technology companies to incorporate into their overall strategies an increased dedication to stable employment, even at some sacrifice in growth rate, market share, or leveraged ROE.

Compensation Policies

Stable employment is, in effect, one kind of compensation policy. What other features should a high-technology company include in its compensation policy? Virtually all top managers of technical companies subscribe to both the laws and the accepted truisms regarding employee compensation policy: no discriminatory practices, emphasis on equity across the organization, and pay based on some combination of merit and seniority. In addition, managers of technical companies need to think carefully about the following, somewhat more subtle, questions related to compensation practices:

- How should the company's wage and salary structure compare to the outside "market"?
- What role should incentive (as contrasted with base) compensation play in total employee compensation? How should the incentives be structured?
- What is the proper mix of deferred compensation and current compensation?
- What is the optimal fringe benefit package?

In considering any aspect of compensation policy, bear in mind that compensation is more likely to be a "dissatisfier" or "demotivator" than a motivator or satisfier. That is, seldom do employees exclaim their satisfaction with a well-thought-out and implemented compensation program. But sloppy or inequitable compensation policies lead to a high level of dissatisfaction and demotivation.

Compensation packages are more likely to dissatisfy than satisfy.

In high-technology fields in which entrepreneurial firms are emerging, compensation policies are sharply influenced by the attraction of these new firms. To hold good managers and technical personnel, existing companies find it necessary to design compensation packages that (1) assure relatively high current income (start-up firms typically require the founding team to suffer some loss of current income), (2) offer stock options or other forms of equity participation (to offset the allure of founders' stock in a new company), and (3) provide some holding mechanism, such as options or deferred compensation, that vests over time (often referred to as "golden handcuffs").

Prevailing Market.   Virtually all companies have a stated policy of paying at or above the prevailing market. Obviously, they cannot all succeed; by definition, half the companies must be paying below the market average.

The definition of *market* presents a problem in implementing whatever policy is finally adopted. Generally, several markets are relevant. The local geographic market is the driving force in setting pay scales for unskilled, semiskilled, and clerical personnel. The market for managerial and marketing talent may be nationwide, and the market for key technical specialists may be worldwide. Regionally-specific and industry-specific wage and salary surveys are available (often at considerable cost) and are essential for monitoring changes in compensation, even if they are not used as the basis for establishing pay ranges. Overreliance on a single survey by all members of an industry, most of whom seek to pay "above market," can result in rapid wage and salary escalation in that industry.

Think carefully about the implications of paying above the market.

Paying above market will lead to above-average costs and below-average margins *unless* the pay policy results in the hiring of truly more effective and efficient employees. Paying above average for average employees is a road to disaster. The implications for the personnel department and all management levels of an above average pay policy are serious; the policy ought not be adopted lightly.

Incentive Compensation.   Incentive compensation in cash is widespread in high-technology companies. There is, however, much disagreement as to the motivational effectiveness of such compensation plans. The traditional view that employees are directly motivated to work both harder and smarter by the promise of higher cash compensation is increasingly questioned. Direct financial motivation is really effective only for persons living at or near the poverty line; note that "piece work" pay practices are limited to industries populated with unskilled or semiskilled employees. Thus, there is increasing doubt that professional employees—knowledge workers—are driven in any useful way by the promise of cash payoffs.

Do cash incentive compensations motivate knowledge workers?

Without delving too deeply into the psychological underpinnings of this argument, one can see that professionals in high-technology companies are motivated by a complex set of factors, a set typically dominated by self-fulfillment goals. Pride, career advancement, and self-actualization (realizing one's potential) play substantially more important roles than does cash compensation.

Cash compensation has two distinct dimensions. First, it represents a form of recognition, some tangible evidence of success (at least, success as measured by the bonus plan). Second, it provides the wherewithal for the recipient to increase his or her consumption or savings. The first dimension is typically more important. This fact may, however, become obscured if organizational norms are that recognition comes primarily in the form of cash bonus payments (that is, the norms reinforce the importance of the bonus plan). Most bonus plans are established with the thought that the second dimension—increased consumption or savings—is the driving motivator. If, indeed, recognition and evidence of success are the motivators, the cash incentive system may be both wasteful of company resources and less effective than other schemes for acknowledging employees' competencies and contributions.

Nevertheless, there are some corporate benefits to be derived from cash bonus (incentive) systems. First, if the bonus comprises a major portion of managers' compensation when averaged across both good and bad years, this portion of compensation becomes a variable cost to the company. That is,

But cash bonus plans are useful communication tools.

high bonuses are paid in exactly those years when the company can best afford them—years of high profit—and total compensation costs are low when the company's financial fortunes are down. Second, a broad-based cash bonus plan—for example, a cash profit sharing plan—serves a useful communication purpose: Participating employees develop a greater sense of community and better identify with the company's financial fortunes. Finally, a cash bonus plan can serve as a signaling device. Depending upon how it is structured, the plan can emphasize growth, return on assets, sales of a new product line, cash flow, or other results to which management wishes to assign high priority.

There are about as many bonus or profit sharing schemes as there are companies employing them. I have categorized the schemes as follows:

1. Companywide profit sharing
2. Group bonus or profit sharing
3. Individual bonus
   - Formula bonus
   - Discretionary award

Profit sharing payments are expected, not a bonus, at successful companies.

Companywide profit sharing is typically set at some percentage (often 10 percent) of pretax profits. In stable and successful companies, such as Hewlett-Packard and Eastman Kodak, cash profit sharing payments are both significantly large and expected by the employees. In less-successful companies, the profit-sharing payments may be trivial; and when profits vary sharply from year to year, the absence of a profit sharing payment in a poor profit year can be a serious demotivator. Thus, these plans have the advantages and disadvantages previously enumerated.

Group bonus or profit sharing may be based upon the performance of a particular division, plant, or sales region. Such plans can be highly divisive, particularly if much cooperation and coordination are required between groups that are subject to different bonus plans. "Game playing" frequently becomes widespread, and rather than focusing on the best interests of the entire organization, employees identify solely with the goals and interests of their bonus-earning groups. Such group bonuses should be avoided except in very large, multiindustry, highly divisionalized companies where the requirement for interdivisional cooperation is minimal.

Group profit sharing plans can be divisive.

Individual bonuses are typically for top executives and sales personnel. These individual bonuses can be further categorized as formula based or discretionary. In many cases, the total bonus is the sum of an amount determined by formula and a discretionary award.

Individual bonuses can be formula based or discretionary.

Most sales incentive plans are formula based. The simplest ones call for payment to the salesperson of a certain percentage of the sale value. In other cases, the percentage varies by product line and may also depend upon the gross margin of the product sold and whether a threshold level of sales has been achieved. (The percentage typically escalates once the threshold has been exceeded.) Such bonus plans have become almost standard among high-technology companies, although some companies, such as Digital Equipment Corporation, compensate their sales personnel solely with straight salary.

Companies that use the plans offer two arguments for the necessity of sales incentive plans. First, the salesperson is operating without close supervision and must be given an "entrepreneurial" payoff (bonus) to assure his or her continuing extra effort. Second, the type of personality that is drawn to a sales position is likely to be financially motivated. Although the validity of these arguments is certainly open to debate, accepted practice essentially dictates the use of sales incentive plans.

However, these plans are not easy to administer. A bonus plan conceived as a simple formula with few qualifiers seldom remains that way. How are sales representatives to be compensated when the sale occurs in one region and the installation and follow-up work are to be accomplished in another region? If price concessions are offered, is the salesperson's bonus affected? Are bonuses paid quarterly, semiannually, or annually? How does management ensure against unreasonable acceleration or delay of orders at the instigation of the salesperson in an attempt to "game" the bonus plan?

Management bonus plans tend to over-emphasize short-term results.

Formula-based bonuses to top executives have many of the same problems. How are both long-term and short-term corporate objectives appropriately acknowledged in the bonus plan? (They probably cannot be; most bonus plans are unreasonably oriented to short-term results.) Is management to benefit from windfall fortunate events and suffer from unforeseen or uncontrollable negative events? (Although set up as a two-way street, too many bonus plans become one-way streets: The bonus formula is altered when external events are judged to be the primary cause of poor results.)

In their frustration to construct both equitable and long-term-oriented formula bonus plans, many top executives and boards of directors opt instead for the awarding of discretionary bonuses. There are at least three problems with such bonus plans. First, they are subjective. Second, they may lead to more posturing than truly effective work—for example, long working hours rather than smart working habits. Third, if awarded on an annual basis, they tend to be geared to performance in the later part of the year because those events remain fresh in the minds of those determining the amount of the award.

Elaborate bonus plans are too frequently pursued as a substitute for careful and thoughtful management. Knowledge workers do not work for the paycheck alone; thus, the "carrot" approach to motivation has limited effectiveness. If the firm shares its financial successes with its employees, there will be many tangible and intangible benefits, but it is unlikely that the motivational effects will be significant. Moreover, individual and group bonuses, whether formula based or discretionary, are very likely to lead to actions that are counterproductive to the corporation as a whole, although beneficial to the individual's pocketbook. The counterproductivity is particularly serious in high-technology companies, where cross-boundary communication and cooperation is so essential.

Bonus plans rarely improve employee motivation.

Deferred Compensation. High-technology companies offer many forms of deferred compensation, the most important of which are typically stock options, stock purchase plans, deferred profit sharing (sometimes in the form of an employee stock ownership trust), and pensions. In areas where competition for key management employees is strongly influenced by new, start-up companies offering equity participation to attract senior managers and engineers, deferred compensation—particularly stock options—is likely to take on increased importance.

Thus, younger companies should emphasize stock options, a form of compensation that has no cash cost to the company (although a very real cost in shareholder dilution). These options have both symbolic and financial consequences. It is amazing how frequently present or prospective employees of a high-technology firm place unrealistic importance and value on a modest stock option award. The option's importance lies in the fact that it recognizes the employee as key. In fact, many more stock options have lapsed or been forfeited than have ever been exercised. Nevertheless, a few individuals have become very wealthy by exercising stock options; so their glamour lives on. The personal tax and other legal ramifications arising from all types of stock options deserve very careful attention, both before the company adopts a plan and before the individual accepts or exercises any option.

Stock options offer both symbolic and financial rewards.

Stock purchase plans have two important benefits. First, they shift the employee's focus more to the company's long-term success—that is, they mitigate the excessive short-term orientation that seems to be a part of most cash bonus or profit sharing plans. Second, they can raise substantial additional capital for the company, a particularly worthwhile benefit if the company's growth rate is expected to outstrip its ability to generate funds internally.

Stock purchase plans fit high-growth, publicly traded companies.

Two important negatives of stock purchase plans should also be recognized. First, employees are led to invest both their careers and their savings in the company; the resulting lack of diversification is risky. Second, the stock's value over the years will swing, depending as much or more upon the vagaries of the stock market as upon the success of the firm. Sophisticated and experienced investors understand this fact, but many employees do not. On balance, stock purchase plans should be strongly considered, but only after the company's stock is publicly traded and a reasonably strong market for the securities is assured.

Mature technology-based companies use pensions much more than young companies. Young companies are likely to be populated with young employees to whom pension benefits seem remote: They will not realize tangible benefits for perhaps thirty or forty years—until they retire. For such companies, the current cost of funding the pensions greatly outweighs the nominal value that the young employees attach to the pension plan. Of course, most such companies, if ultimately successful, must someday institute a pension plan, and, at that time, they will face a substantial unfunded pension liability.

Young employees attach little value to pension plans.

Deferred profit sharing plans are a useful substitute for pension plans. The benefits are still deferred, but the company's ultimate responsibility to provide long-term savings and retirement benefits to the employee is recognized. Furthermore, company contributions to the plan are geared to profits, with the result that contributions are scaled over the years to the company's ability to pay.

Get legal and tax advice before offering any deferred compensation plan.

All deferred compensation plans are circumscribed by elaborate and confining federal legislation, most of which, unfortunately, has become necessary over the years to eliminate abuses. The tax impact to both the individual employee and to the company can be dramatic. Seek expert legal and tax advice in the early stages of designing any deferred compensation plan.

You can't attract and keep good employees without attractive fringe benefits.

Other Fringe Benefits. Technical companies also offer many other benefits to their employees. Today, fringe packages include, as a minimum, vacations (at least two weeks, and increasingly three and four weeks, particularly after several years of continuous employment); group medical, life, and long-term disability insurance plans; and a reasonably full complement of holidays (although a modest number by European standards). To attempt to attract and retain competent employees without this minimum set of fringe benefits is hopeless.

Technical companies have become increasingly ingenious in offering other attractive fringe benefits, particularly in those geographic regions where competition for good technical personnel is extreme. Some of these benefits, which tend to be especially attractive to knowledge workers, are the following:

1. Educational benefits. Today, most technical companies pay for job-related continuing education, particularly if the course work is leading to an advanced degree. Some companies make this benefit broadly available; others provide it as a privilege only to those selected. The benefit to the company of encouraging additional education of its employees is obvious.

2. Recreational facilities. With increased interest in health, employees are being attracted by locker rooms, workout rooms, and, increasingly, tennis courts, swimming pools, and fully equipped gymnasiums.

3. Sabbaticals. Adopting a practice of the universities, a number of technical companies are now offering their employees a certain number of months (typically, from one to six) of paid sabbatical leave every so many (typically five to ten) years. Such leaves are expected to provide regeneration of the knowledge worker, although most companies offering sabbaticals do not insist that activities during the leave be in any way job or technology related.

Fringe benefits will undoubtedly continue to grow as a percentage of the total compensation package paid by technical companies. Fringe costs are still modest in this country compared to other industrialized countries. The aggressive and fast-growing companies—those that seek to increase their employment ranks rapidly—will continue to lead the way toward more elaborate, and undoubtedly more expensive, fringe benefits. Each technical company needs to make a conscious decision whether it will be a leader or a follower in offering fringe benefits. Whether leader or follower, the company must expect to add new benefits periodically in order to maintain its desired posture in the competitive market for knowledge workers.

Should you lead or follow in the race to add fringe benefits?

∴ Personnel policies in high-technology companies must be different from those in companies employing fewer knowledge workers and experiencing less rapid change. Matrix organizations are essential, as are dual ladders of promotion, compensation, and recognition for technical and management personnel. Stable employment levels are both beneficial and difficult to achieve in environments characterized by rapid and abrupt change. Compensation policies should stress deferred compensation and fringe benefits that appeal particularly to knowledge workers. Because incentive compensation plans are difficult to administer and frequently prove counterproductive, do not view them as a panacea for the difficult motivation and compensation challenges that face knowledge-intensive businesses.

Highlights

- In high-technology companies, general managers must both do and manage; the former predominates in small and new firms and the latter in mature and large firms.

- General managers, whose jobs combine leading, performing ceremonial tasks, and integrating company functions, must be action oriented, flexible, and able to live with ambiguity.

- As a company grows and progresses through stages, the demands on managers change, requiring adaptations to both organizational structure and management style.

- Top managers in high-technology firms must be conversant with their companies' technologies and should have held positions in several of its key functional areas.

- A corporate culture (the beliefs, values, and policies that shape and are shaped by business strategy) facilitates communication and decision making, but takes time to build and is difficult to alter.

- In order to maintain and enhance creativity in high-technology companies, managers must be sure that change is accepted, balance short- and long-term objectives, tolerate some bootlegging, foster and reward risk taking, encourage champions of new ideas, and build balanced technical staffs.

- Matrix structures and dual ladders of promotion, compensation, and recognition for technical and management personnel are essential in high-technology firms.

- Although difficult to achieve in environments characterized by rapid and abrupt change, stable employment levels benefit both the business and its employees.

- Incentive compensation plans are troublesome to administer and often prove counterproductive in knowledge-intensive businesses.

Notes

1. Simon Ramo, *The Management of Innovative Technological Corporations* (New York: Wiley-Interscience, 1980), p. 157.

2. Steven C. Brandt, *Strategic Planning in Emerging Companies* (Reading, Mass.: Addison-Wesley, 1981), p. 80.

3. H. Edward Wrapp, "Good Managers Don't Make Policy Decisions," *Harvard Business Review* (September–October 1967), pp. 91–99.

4. I first heard this phrase from Dr. John Doyle, currently vice-president of research and development at Hewlett-Packard Company, and have always credited him with its invention, although subsequently I have seen and heard the phrase elsewhere.

5. Thomas J. Peters and Robert H. Waterman, *In Search of Excellence* (New York: Harper & Row, 1982), chap. 5.

6. Modesto A. Maidique and Robert H. Hayes, "The Art of High Technology Management," unpublished working paper, Stanford University, May 1982.

7. Larry E. Greiner, "Evolution and Revolution as Organizations Grow," *Harvard Business Review* (July–August 1972), pp. 37–46.

8. Ibid., p. 43.

9. Ibid.

10. Peter F. Drucker, *Management: Tasks, Responsibilities, Practices* (New York: Harper & Row, 1973), pp. 32–33 and 176–177.

11. Edward B. Roberts, "Generating Effective Corporate Innovation," *Innov-aha!-tion*, pamphlet prepared by the editors of *Technology Review*, pp. 3–9.

Chapter Nine

Formulating an Integrated Strategy in High-Technology Companies

This chapter covers the following topics:

- Defining Strategic Planning: The Emphasis on Change
- Picturing Strategy in a High-Technology Business: Trade-Offs, Coherence, and Integration
- Using the Strategy Diagram: Three Generic Strategies
- Assessing Competitive Position: Your Firm and Its Rivals
- Guidelines for Successful Strategic Planning

Strategy is not a new topic in this book. I have discussed the pieces of strategy: engineering, marketing, production, finance, and human resources. In this final chapter, drawing on these earlier discussions, I put the pieces together. The task at hand is to formulate, integrate, and implement a coherent strategy within a high-technology company.

I begin by reviewing the objectives of and the forces impinging upon business strategy. Then I outline a procedure for reducing the almost overwhelming question of "What should the strategy be?" into more manageable subquestions. By addressing these questions, strategists not only make their job more tractable, but they also can better assess the broad range of strategic opportunities open to the business, assure that the resulting plan is integrated (that they have not overlooked some issue or condition that could undermine the strategy), and test the strategy to be certain that it is consistent with competitive realities and prospects.

Not only must the high-technology company's strategy be both internally well integrated and consistent with the external competitive environment, it also must be dynamic and evolutionary. Rapid change is a certain condition of high-technology industries; so I conclude this chapter—and the book—with a set of guidelines for successful strategic planning in a fast-moving environment.

A key premise throughout this discussion is that overall business strategy both drives and is defined by the actions and substrategies pursued by each function of the business: engineering, marketing, sales, production, finance, and human resources. I have emphasized repeatedly that the complexity of technology-based businesses requires that management focus particular attention on the interfaces between the primary business functions. Necessarily, then, formulating—and, from time to time, reformulating—strategy within a technical company requires careful orchestration. Top managers must be certain that middle managers understand the overall business strategy and that each functional area's actions and tactics are both consistent with and supportive of that overall business strategy.

Defining Strategic Planning: The Emphasis on Change

Strategic planning leads to commitment of resources and action.

Peter Drucker frames his definition of strategic planning by contrasting what it is and is not:

1. It is not a box of tricks, a bundle of techniques. It is analytical thinking and commitment of resources to action.
2. Strategy planning is not forecasting. It is not mastermminding the future.... Forecasting attempts to find the most probable course of events or, at best, a range of probabilities.... The central entrepreneurial contribution, which alone is rewarded with a profit, is to bring about the unique event or innovation that changes the economic, social, or political situation.

3. Strategic planning does not deal with future decisions. It deals with the futurity of present decisions. . . . The question is "What do we have to do today to be ready for an uncertain tomorrow?" . . . We can make decisions only in the present and yet we cannot make decisions *for* the present alone.

4. Strategic planning is not an attempt to eliminate risk. It is not even an attempt to minimize risk. To take risks is the essence of economic activity. . . . It is essential that the risks taken be the right risks. The end result of successful strategic planning must be capacity to take a greater risk. . . . We must understand the risks we take.[1]

Strategic planning in a high-technology industry implies an innovative strategy that adapts to new opportunities arising from the markets and from evolving technology and to present or potential threats from competitors. Drucker goes on to highlight characteristics of an innovative strategy:

> The ruling assumption of the innovative strategy is that whatever exists is aging. The assumption must be that existing product lines and services, existing markets and distribution channels, existing technologies and processes will sooner or later—and usually sooner—go down rather than up. The governing device of a strategy for the ongoing business might therefore be said to be: "Better and More." For the innovative strategy the device has to be: "New and Different." . . . Systematic abandonment of yesterday alone can free the resources and especially the scarcest resource of them all, capable people, for work on the new.[2]

New and different not just better and more.

Strategic Planners

Our viewpoint in this chapter is that of the business unit general manager. Nevertheless, planning and implementing strategy is by no means solely his or her responsibility. Business strategy both drives and is driven by functional strategies: the set of objectives, priorities, and competencies of engineering, manufacturing, sales, marketing, finance, and personnel. These objectives and priorities are often bound up in the organizational philosophy and values—the corporate culture. Where it is strong, corporate culture both guides strategy formulation and provides a certain inertia against radical changes in strategy. As we shall see, a certain amount of inertia is probably, on balance, beneficial.

All key functional managers must participate in strategic planning.

All key functional managers must participate in strategy formulation and be active as change agents in implementing that strategy. If they are not, they will neither understand nor subscribe fully to the business strategy, and the result will typically be a static, nonintegrated, or incompletely implemented strategy.

Business Versus Corporate Strategy

I am approaching strategy from the perspective of the general manager rather than the corporate CEO. In a small company, they may be one and the same: The general manager is the president-CEO. But in a large company, the general manager whose viewpoint we assume is likely to carry the title of division manager or unit manager.

Corporate (as contrasted with business) strategists tend to view a multiindustry, multinational corporation as a "portfolio" of business units, each with different strategies, opportunities, and risk profiles. They seek a balanced portfolio of businesses—some growing fast and thus requiring large investments and others more mature and generating cash. The CEOs of Fortune 500 companies are properly portfolio strategists. But our attention is on the operating general manager who is responsible for individual business unit strategy.

Different corporate divisions require different business strategies.

Within a single large company, individual business units may have different strategies. Rockwell International employs distinctly different strategies in its Graphic Systems Division, a manufacturer of printing presses, and in its North American Aviation Division, a defense contractor. Occidental Petroleum has a different strategy for its Hooker Chemical operations than for its oil or meat processing operations. Within Hooker Chemical, the strategy of the Zoecon Division, producing state-of-the-art pharmaceuticals, bears little similarity to that of the industrial chemical divisions. Even within companies typically characterized as wholly high technology, different strategies may be required across the spectrum of operations. HP's approach to the computer and related data products business is a good deal different than its approach to the medical electronics or test-and-measurement instruments markets, and TI needs different strategies for its semiconductor operations and its defense contracting divisions.

Excessive Diversification

An exciting and successful business strategy in a high-technology company is more likely to demand focus and concentration than diversification. High-technology firms that have diversified outside of high technology have generally been less successful than their counterparts that have concentrated their attention on related markets, products, and technologies. The conglomerate strategy, all the rage in the 1960s, has now been generally discredited. Managers of large corporations whose activities range across, for example, high- and low-technology manufacturing, financial services, construction, and retailing have difficulty remaining fully abreast of threats and opportunities in each business arena. Moreover, to oversee and to allocate financial resources among operations with very different risk profiles requires an unusually able management team. Most such managements tend to shortchange their high-technology operations in favor of operations that are easier to understand and involve less risk.

Focus is generally more effective than diversification.

Excessive diversification is typically undertaken either as a financial manipulation (to improve earnings per share through increased debt leverage or pooling of interest accounting) or a desperate redeployment of assets and management attention away from present businesses that have fallen behind competitively and are therefore inadequately profitable. David Packard, co-founder of HP, summarizes the point well: "More businesses have died of indigestion than of starvation."

Diversification should be resisted, but not avoided altogether. The diversity must have some coherence. For most companies, that coherence is built around a common *market*. For high-technology companies, the dominant common thread may be technology, but ideally it will be *both* technology and markets.

Diversify coherently through both technology and markets.

∴ Strategic planning in high-technology businesses requires innovation and adaptability. If it is to be successful, it must be done by all key functional managers and must include an assessment of the impact of both technology and market needs.

 Picturing Strategy in a High-Technology Business: Trade-Offs, Coherence, and Integration

Discussions of strategy seem inevitably to involve diagrams, tables, or graphs to represent the strategic elements and choices. This is not surprising because formulating business strategy is a matter of making choices among alternatives and balancing the many internal and external forces that impinge on that strategy. Diagrams are particularly useful in illustrating the messiness, as well as the trade-offs, inherent in building a coherent business strategy.

The interaction of markets and technology creates opportunities and threats.

Figure 9-1 draws together the various discussions in this book in a representation of the business strategy of high-technology companies. What distinguishes a high-technology business from its more mundane cousins is that technology is a key ingredient in the company's strategic plan. A key distinguishing feature of this figure is that it shows technology and markets as the major forces that shape strategy. A kind of creative tension can and should exist between these two forces: Technology both affects markets and is affected by them. In businesses less dependent on technology, strategy is derived largely from the market alone. In high-technology businesses, viewing either markets or technology as the dominant source of strategy is dangerous. The key is to focus on the interaction

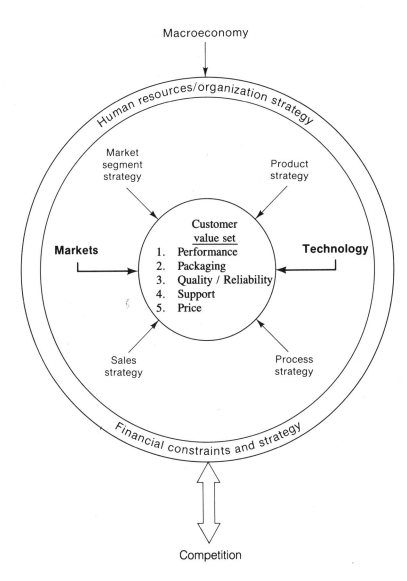

Figure 9-1. Business strategy in high-technology companies

between technology and markets because that interaction creates both strategic opportunities and competitive threats.

Indeed, this is where I started the discussion in Chapter 2. I emphasized there that a—and arguably *the*—key job for the general management team is to select those product-market opportunities the technology-based company should pursue. Myopia with respect to either markets (the sales executive's tendency) or technology (the engineer's tendency) will result in a suboptimal strategy. Linking emerging technology and emerging market needs can result in a very powerful business strategy; one without the other can lead to enormous financial drain. General managers, in their role as strategists, must be thoroughly knowledgeable in both the technology and the markets. Too many top managers, in their discomfort with sophisticated technology, attempt to delegate the responsibility for understanding the company's technological underpinnings. An unbalanced or incomplete business strategy is almost certain to result from this abdication of responsibility.

The Objective

Focus on the values most important to your customers.

The objective of any successful business strategy must be to deliver value to the customer such that the relationship between the price the customer is willing to pay and the cost of delivering the product or service will result in above-average returns (financial and personal) to the owners and employees of the business (see the center of Figure 9-1).

The customer is interested in neither the supplier's strategy nor its technology. The customer is interested in acquiring value. Fortunately, value is multidimensional, and not all customers seek the same mix of values. All, however, are seeking some composite of the following five elements:

1. Performance. The customer is likely to begin his or her appraisal of value with the technical and operating specifications of the product or service. This value dimension is essential—but typically not sufficient—to assure the sale.

Your customers seek a mix of performance, packaging, quality, support, and price values.

2. Packaging—the physical, esthetic, and emotional surroundings of the product or service. Who could deny that the physical packaging, logo, graphics, and image as a "scrappy" young company were important ingredients in Apple Computer's spectacular early success?

3. Quality at the time of delivery and reliability in use over time. The customer seeks optimum, not necessarily maximum, quality and reliability.

4. Support. An increasingly important value dimension for high-technology products and services, support may occur both before and following the sale and may be delivered by various functions of the technical organization, not solely field service.

5. Price (more accurately, price in relation to the four previously listed elements). One of the appeals of high-technology businesses is that price-insensitive product-market segments abound. But they are not exclusively price insensitive; in some segments, price competition dominates.

One of the widely acknowledged attractions of high-technology product-market segments is that elements of the customer value set other than price tend to dominate. The attraction derives from the fact that companies competing in price-insensitive arenas typically achieve above-average profitability. For a product-market segment to be price insensitive, buyers must be particularly sensitive to (that is, motivated by) performance, packaging, support, and/or quality/reliability. To assist the strategist in assessing the price insensitivity of product-market segments, Figure 9-2 lists buyer characteristics that result in price insensitivity and examples of products, systems, or services.

Price-insensitive product-market segments can be particularly profitable.

Strategists should bear in mind that the purchasing decision for many technical products and services is influenced by several decision makers within the firm. When purchasing's influence is high, price is likely to be accorded relatively high priority. When others in the organization are the dominant decision influencers, price is more likely to take a back seat. But it is not enough to hope for or assume price insensitivity simply because the product or service involves sophisticated technology. If the product or service does not fit into one or several of the categories listed in Figure 9-2, you should anticipate and prepare for price competition.

Business Strategy Elements

The task for the high-technology company is to design and deliver an optimum set of values to customers within the context of the following sources of opportunities and constraints:

Performance-sensitive buyers

OEM purchasing products that represent a small cost but are critical to operation (e.g., microprocessors, transducers)

Purchaser of capital equipment promising process cost saving (e.g., process instrumentation, N/C machine tools)

Purchaser seeking custom-engineered product, system, or service (e.g., "turnkey" process plant, material handling systems)

Packaging-sensitive buyers

Buyer who is not knowledgeable about the product (e.g., business computer systems, process control systems for mature industries)

OEM seeking to "trade" on the image and reputation of the component supplier (e.g., IBM Selectric typewriter mechanisms, jet engines)

Support-sensitive buyers

Unsophisticated buyer of capital equipment requiring user training and preventive emergency maintenance (e.g., analytical instruments, word processing systems, medical electronic equipment)

Buyer of customized equipment or systems requiring extensive application engineering (e.g., automated warehousing systems, robotic equipment)

OEM seeking forewarning of design changes, flexibility in delivery, and similar forms of supplier support (e.g., integrated circuits, subcontracted fabrication or assembly)

Quality/reliability-sensitive buyers

OEM acquiring components or subsystems for which field repair is difficult or expensive (e.g., computer software, minicomputers, motors)

Buyer of capital equipment that is central to the buyers' operations (e.g., on-line computer systems, electrical control panels)

Figure 9-2. Buyer characteristics resulting in price insensitivity

1. Markets, those that exist and those that can be developed

2. Technology that exists and technology that can be developed (See Chapter 4 for a discussion of technical policy within the firm.)

3. The financial constraints of the firm and opportunities available to overcome them

4. The macroeconomic environment—national and international—in which the company operates

5. Competition, both existing and potential

There are five key messages in Figure 9-1 for the business strategist. First, the *market* is the dominant determinant of both *market segment* strategy and *sales* strategy. The questions of what markets to attempt to access and how to organize to access them should be considered separately. New market segments are always appealing, but a novel sales approach (for example, via retail stores, mail order, or OEMs) to a market segment that is already well developed may offer a significant strategic advantage.

Analyze sales strategy independent of market segment strategy.

Second, *technology* is the dominant determinant of both *product* and *process* strategies. I give these two strategic elements equal weight in Figure 9-1. U.S. high-technology industries have been slow to capitalize on the opportunities inherent in improved process technology. In the United States, most R&D efforts are product oriented. Japan has demonstrated that conscious development efforts devoted to processes—that is, assigning first-rate engineers to improve manufacturing technology—can yield large payoffs.

Don't overlook competitive leverage derived from process strategy.

Third, the strategist should strongly resist the unfortunate tendency to view marketing and R&D as the only sources of strategic advantage. As I stressed in Chapters 3 and 4, manufacturing has proven to be a key strategic weapon for the Japanese, TI, National Semiconductor, and others. Finance can also be a key strategic weapon. IBM's financial muscle permits the company great competitive advantage in setting lease and sale terms for its customers. Offshore financing or asset *de*-intensification may offer other strategic possibilities. *Human resources* and *organizational* strategy, although somewhat less tangible (or at least more difficult to measure), have clearly been primary strategic elements for 3M, IBM, HP, and others.[3]

Fourth, sales strategy and manufacturing strategy are both inherently process strategies and they should be linked. Both involve issues of forward and backward integration. These strategies have major impacts on the customer values of packaging, support, and price.

In sum, six functional strategies must come together, be mutually consistent and supportive, and thereby both define and encompass the business strategy:

- Market segment
- Sales
- Product
- Process
- Financial
- Human resources and organization

A Dynamic Situation

A shortcoming of Figure 9-1 is that it is a snapshot; a moving picture might be more appropriate. That is, the figure implies relatively static conditions. The true condition is quite the opposite: Not only are markets and technologies changing— sometimes very rapidly—but the economic environment is changing and competitors are both acting and reacting. Thus, the business strategy must be developed and adapted with due attention to the dynamics of both the economy and competitors: the final two arrows.

All but the very largest of companies must accept the macroeconomy as a given condition over which they have little control—a so-called *exogenous variable*. Thus, the arrow on "macroeconomy" on the diagram is one way. Although the timing of economic cycles affects strategic planning, these cycles can seldom be used as a strategic weapon, despite many comments to the contrary. Economic forecasters have been notoriously inaccurate in recent years. Strategists should assess the interplay between various economic scenarios and alternative business strategies, planning explicitly for defensive actions in case of economic weakness and for offensive actions in an economic upturn. But a business plan formulated in late 1981 that counted on economic recovery in 1982 could have been disastrous, in spite of the general concensus among economists in late 1981 that recovery was imminent.

Expect your strategy to influence and be influenced by your competitors.

Competitors must not be taken as a given. Thus, the arrow on "competition" in Figure 9-1 is two way. The firm's business strategy can influence competitors' actions just as surely as competitors' moves will shape, in part, the firm's strategy.

Integration

The overriding message in Figure 9-1 is the importance of integration. A business strategy is a composite of functional strategies. The general manager—that is, the chief business strategist—must be certain that the functional strategy pursued in each area of the business is not incongruent with the overall business strategy. For example, the sales strategy must not involve indirect distribution if the market segment strategy aims at very technically qualified buyers and the product strategy involves customized products. If the market segment strategy calls for heavy emphasis on price within the customer value set, then manufacturing's strategy had better emphasize process improvements and vertical integration and the financial strategy had better permit such investments.

Be certain that functional strategies support the overall business strategy.

Figure 9-1 provides the strategist with a kind of checklist to work through in determining both the possible sources of strategic advantage and all the elements of strategy that must be orchestrated to ensure that the final strategic plan for the business is coherent.

Figure 9-1 also carries an important message for the functional manager (for example, the sales manager or the manufacturing manager). These lower-level managers must *influence* the strategy formulation process by forcefully presenting the constraints and opportunities inherent in their operations, *understand* the overall business strategy that is finally adopted, and then *adapt* their operation to support it. Failure to do so will result in departmental performance that senior managers will consider unsatisfactory. The functional manager who pursues an incompatible departmental strategy will soon be out of a job. Moreover, the functional manager may face a struggle in overcoming senior managers' resistance to those changes and investments that are essential to align department strategy with business strategy.

Functional managers must influence the formulation of strategy and then understand and adapt to it.

∴ The objective of strategic planning is to deliver a mix of performance, packaging, support, quality/reliability, and price to the customer so that the relationship between price and cost will yield above-average returns to the business. Strategic planning is a dynamic process that involves the seizing of appropriate, although risky, opportunities. It both drives and is driven by functional strategies; so it involves the integration of market segment, sales, product, process, financial, human resources, and organization substrategies.

Using the Strategy Diagram: Three Generic Strategies

We now have a framework to use in developing a business strategy and assessing its effectiveness (customer value versus cost) and its coherence across the functions of the high-technology company. I illustrate the use of this framework by considering three widely recognized—and therefore in a sense generic—strategies. In the course of the discussion that follows, refer frequently to Figure 9-1 and note particularly (1) the explicit consideration of the mix of values the customers seek (the three strategies assume different priorities among the elements of the customer value set) and (2) the need to consider each of the functional strategies both independently and in the context of the strategies of its sister functions—that is, the need to orchestrate the functional strategies.

Many writers on business strategy have identified these three fundamental or generic strategies. Porter labels and illustrates them as shown in Figure 9-3:

The three fundamental strategies are overall cost leadership, differentiation, and niche.

1. Overall cost leadership: high-volume, attractively priced, standardized products sold to broad market segments (such companies as Sony, TI, Osborne Computer, Epsom, and Honda)
2. Niche: unique (often custom or customized) products and services, typically emphasizing the nonprice values of per-

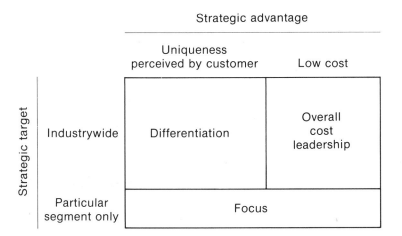

Figure 9-3. Three generic strategies
Source: Michael E. Porter, *Competitive Strategy* (New York: Free Press, 1980), p. 39.

formance and packaging, sold to market segments of limited scope (many small technical companies, including most—but certainly not all—start-up companies; medium-sized companies include Spectra-Physics, Finnigan, California Microwave, Scientific Atlanta, and Rolls Royce)

3. Differentiation: products and services offering unique non-price values (typically performance, support, and quality/reliability) to broad markets (such companies as IBM, HP, Bechtel, Boeing, Syntex, and Mercedes Benz)[4]

Other authors use different names for essentially the same three strategies. Overall cost leadership may be referred to as "low-cost producer" or "commodity house" (because truly low cost necessitates product standardization or commoditization). Porter's "focus" strategy is more typically referred to as a "niche" strategy and sometimes as a "boutique" strategy.

Low-Cost Producer

Here customers emphasize price, quality, and performance among the possible set of values and they demand little support. A business pursuing this strategy needs to accumulate

The low-cost producer strategy implies broad markets, standardized products, and an emphasis on process strategy.

experience in excess of its competitors, and thus the low-cost producer strategy implies the following functional strategies:

1. A *market segment strategy* that targets broad, probably industrywide and possibly worldwide, markets

2. A *sales strategy* that is consistent with capturing and maintaining a high (preferably dominant) market share and that offers only limited support to the customer

3. A *product strategy* that emphasizes standardized products and ones that are engineered for automated (or at least easy) manufacture (Product improvements are likely to be incremental rather than revolutionary, first, because customers place higher priority on price and reliability than on performance and, second, because revolutionary product changes would interrupt steady progression down the experience curve.)

4. *Process strategy* that ensures a steep slope for the experience curve: automated operations, a strong emphasis on manufacturing engineering, backward integration, rapid inventory turnover, and a quality level that minimizes warranty repairs

5. A *financial strategy* that supports the heavy investment in fixed assets that will probably attend the process strategy and pays close attention to asset management and cost control

6. *Human resources and organization strategy* that emphasizes efficiency, stable employment, and routinization of cross-functional relationships

Focus or Niche

The niche strategy implies direct sales, much customer support, unique products, and flexible processing.

The customers are typically price insensitive and are likely to emphasize performance (against a specific rather than a broad set of requirements), packaging, and support among the possible set of customer values. The niche business strategy implies the following functional strategies:

1. A *market segment strategy* that focuses on small segments not being well served by products or services that are broadly available from large-volume suppliers. The segments are restricted by geography, buyer group, a limited product

line, or a combination of these three. Depending upon its size, the company seeks a single niche or a series of niches, but resists the temptation to tackle broad markets.

2. A *sales strategy* that calls for a highly trained and technically sophisticated sales force, often with heavy emphasis on application engineering. Support typically represents a significant competitive advantage and is delivered by all of the sales force, the application engineers, and the field service engineers.

3. A *product strategy* at the opposite end of the spectrum from the low-cost producer strategy. Unique products tailored to specific applications are stressed. Product customization is also important in some niches. The highly focused products deliver such value to customers that high prices can be obtained, permitting, in turn, heavy expenditures on development, sales and marketing, and low-volume manufacture.

4. *Process strategy* that gives high priority to flexibility in order to adapt to new technology and changing customer requirements. Processes are likely to be general purpose and labor intensive, and backward integration is generally disadvantageous.

5. A *financial strategy* that reflects low fixed asset intensity. Working capital investments are likely to be high because receivable collection periods are typically long (customers evaluate actual performance before authorizing payment) and inventory turnover is slow (low parts usage rate, long in-process time, large spare parts stock to serve customers well). Demanding down payments from customers can mitigate this condition. High ROS can offset low debt leverage and low working capital turnover to result in above-average ROE.

6. A *human resources and organization strategy* that emphasizes creativity, minimum structure, informality, maximum cross-functional communication, and equity incentives within the compensation package.

Differentiation

The differentiation strategy is similar to the niche strategy in seeking competitive advantage through product or service dif-

The differentiation strategy implies broad markets but unique, well-supported, high-performance products sold direct.

ferentiation rather than price and similar to the low-cost producer strategy in seeking to serve broad rather than narrow markets. Performance, support, and quality/reliability dominate the customer value set. The differentiation business strategy implies the following functional strategies:

1. A *market segment strategy* that targets broad markets, as with the low-cost producer strategy, but markets whose buyers perceive and value performance, support, and quality differences more highly than price differences.

2. A *sales strategy* that calls for heavy expenditures on marketing communication and selling in order to reinforce the company's competitive distinctiveness. Direct selling is likely, and application engineering and field service may again be key in delivering support value to the customers.

3. A *product strategy* that is somewhere between the niche and low-cost producer strategies. The products and services are likely to be complex, comprise a broad line (having a wide range of capabilities), and frequently will involve a system—that is, a combination of hardware and software elements. New product opportunities may arise from a broadening as well as a deepening of internal technical capabilities. The product and sales strategy together will seek to develop a high degree of customer dependency on the firm; this dependency can translate into high prices and repeat (or add-on) sales.

4. *Process strategy* that reflects product complexity and uniqueness. Close communication between engineering and both marketing and production is necessary. Processes facilitate hardware and software integration and are well meshed with the support services the sales and field service departments deliver. Vertical integration is sought as a source of differentiation more than as a route to lower costs.

5. *Financial strategy* that permits the inherently high asset intensity. The firm's cost structure is dominated by managed or discretionary expenses (development and marketing), and thus budgeting assumes great importance. If above-average ROE is to be achieved, high ROS must be derived from price premiums (more than low operating costs) in order to offset the low asset turnover. Direct or indirect

financing of customers can be used as another competitive weapon.

6. A *human resources and organization strategy* similar to the niche strategy in that it stresses creativity and cross-functional communication and coordination. Delegation is essential, and a formal or informal matrix organization is likely. Task forces may be useful. Dual ladders of promotion and compensation are common, and attractive fringe benefit packages are likely to be important in attracting highly skilled engineering, marketing, and manufacturing personnel.

High Cost of Incoherence

A company that does not establish a clear-cut business strategy and then align its functional strategies with that overall strategy is likely to find itself caught in a no-man's land. For example, if the market segment and product strategies are targeted at standardized products sold to large markets, but the process and sales strategies are appropriate for the niche or differentiation strategy, the result will be high cost, low asset turnover, and a distressingly poor ROE. A company whose process and financial strategies are appropriate for a low-cost producer, but whose market segment and sales strategies are targeting a niche opportunity, will realize frustratingly low growth and be unable to capitalize on the operating and debt leverage that it has acquired.

An incoherent strategy results in poor profit performance.

Benefits of Stubborn Consistency

The forces of change that are inherent in high-technology industries frequently cause niche markets to broaden and well-developed products and markets to become standardized and increasingly price sensitive. Specialized products tend to become standardized over time, and, of course, small markets have the happy tendency of growing into large markets. How should the high-technology company respond to these changes?

When a niche market reaches reasonable size, the company's strategy must either take on many of the characteris-

Resist changes in your fundamental strategy as markets expand and products standardize.

tics of the differentiation strategy or seek other niche markets. When the products and services of the company pursuing a differentiation strategy become mature and somewhat more standardized, the firm must either adopt certain of the functional strategies of the low-cost producer or take action to reestablish differentiation. Too many high-technology companies choose the first rather than the second course of action. The result is that they slide into an incomplete, ineffective, and unprofitable change in strategy as they are seduced into pursuing broader markets with more standardized, price-sensitive products.

Changes in strategy—from niche to differentiation or differentiation to low-cost producer, for example—can and should be resisted in most instances. As the original niche market grows substantially, subniches almost always appear. The competitor particularly capable of and experienced in implementing the niche strategy should stick with it, seeking out these subniches. Such a company must include as part of its niche strategy conscious actions to create new specialized products or services that cater to niche (or, if you will, subniche) markets. As the differentiated product or system matures and becomes more standardized, the competitor particularly adept at the differentiation strategy should set about in its development and product planning efforts to reestablish differentiation rather than to compete on unfamiliar ground as a low-cost producer.

Seek new niches and re-create differentiation.

The systematic abandonment of existing products and market segments in favor of new niche and differentiation opportunities is necessary. The consistent—even stubborn—pursuit of a single strategy is likely to be substantially more successful and profitable than a shift to the low-cost producer strategy that entails an overhaul of the company's policies and procedures with attendant organizational upheaval.

∴ Use of the strategy diagram is illustrated utilizing three generic strategies—low-cost producer, niche, and differentiation. Each implies a different mix of customer values and must be supported by different functional strategies. Stubborn adherence to a single strategy is generally more successful than pursuit of a hybrid strategy.

Assessing Competitive Position: Your Firm and Its Rivals

One cannot speak of business strategy in isolation, but only in the context of competitors' actions and reactions, at least when considering nonmonopolies in free-enterprise economies. Thus, Figure 9-1 shows a two-way arrow on "competition." To this point, I have focused primarily on the internal capabilities, operations, policies, and priorities of each of the functional areas of the business and on customers. How can the strategist go about assessing the *relative* position of competitors and his or her own company?

Consider business strategy within the context of your competitors.

Writers on business and corporate strategy have been prolific in suggesting frameworks (typically matrices) to use in assessing relative competitive positions within a product-market segment. Most such frameworks focus on the interaction between market attractiveness and market share. In high-technology industries, the various competitors' technical prowess must also be factored into the equation.

Figures 9-4 and 9-5 present two simple matrices for assessing competitive position, one focusing on technology and the other on markets—the two main forces in Figure 9-1. The concepts behind the first matrix, and particularly the colorful names in each box, are generally credited to the Boston Consulting Group. The technology matrix was developed by Booz-Allen and Hamilton consultants.

Consider your relative competitive position in both market share and technological competence.

In each figure, the X-axis calls for an assessment of the company's position not on an absolute scale, but relative to its competitors. Thus, a number two position in an industry might still leave the company in a weak competitive position if the number one company has a commanding lead in either market share or technology.

The Y-axis focuses on market and technology attractiveness. A fast-growing market provides a more attractive competitive arena than a mature or declining market; similarly, a high-technology company seeks to compete where its technological strengths are important.

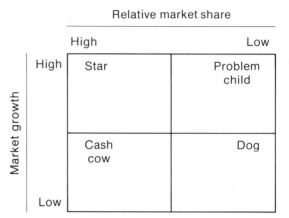

Figure 9-4. Market matrix

The names within the cells of the matrices are similar. The "high-high" cell offers the real opportunity for the future. A large share of a high-growth market presents a "star" opportunity for company growth and profitability. A commanding technological lead in a market where technology is important represents an opportunity in which to invest resources or "bet." The opposite—the "low-low" cell—represents a business from which withdrawal should be sought, gracefully, if possible.

Commanding market or technology leads in slow-growth markets, or markets no longer dependent upon technology, are desirable because such businesses produce cash. They are typically profitable and, because of slow growth and mature technology, demand little reinvestment in either facilities or development. But these cells do not present opportunities for growth and new investment.

The final cell in each matrix presents a particular dilemma to management. The company finds itself in a weak competitive position—in terms of share or technological capability—within an attractive industry, one that is growing fast or in which technological leverage is possible. Appropriate action depends, in part, upon what alternative opportunities are available to the company. To become an effective and profitable competitor in this troublesome sector will require heavy investment in building either market share or technical com-

Weak positions in attractive markets and technologies are the real dilemmas.

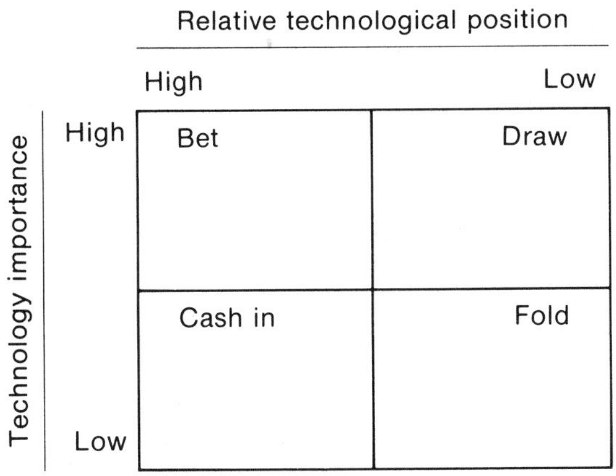

Relative technological position

	High	Low
High	Bet	Draw
Low	Cash in	Fold

Technology importance

Figure 9-5. Technology matrix

petence. If the company has several "star" or "bet" opportunities within its portfolio of product-market segments, it probably has a full agenda of investment opportunities and should withdraw from the "problem child" segment. If it has a portfolio full of "cash cows" and no "stars" or "bets," it may decide to invest in its "problem children" in an attempt to move them into the "star" category.

∴ High-technology companies should assess their position relative to competitors in terms of both technological capabilities and market share. They must do so in light of market growth and opportunities for exercising technical leverage.

◆ *Guidelines for Successful Strategic Planning*

Strategic planning is hard work. I conclude this chapter by offering some guidelines to help you accomplish it successfully. Understanding common causes of poor planning, as well

Strategic planning is hard work.

as the primary sources of risks and opportunities, can improve your planning effectiveness.

Fundamental Questions

Strategic planning is fundamental to all operating and financial planning.

Why?   The strategic plan is fundamental to all other organizational plans. Long-range personnel planning, capital budgeting, profit planning, and all other short- and long-term plans must be developed in light of the business's strategic plan. The strategic plan is key to communication, particularly interfunctional communication. Ideally, the essential and lasting elements of the strategic plan—for example, basic precepts of financing and human resources management—become built into the corporate culture.

Involve all functional executives in strategic planning.

Who?   The classic mistake is to assign the task of strategic planning to a single top executive—a "planner"—or, worse yet, to a committee. A planning executive, committee, or department should play no greater role than that of data gatherer, facilitator, and coordinator. Just as budgeting departments should not budget, planning departments should not plan. Line managers must plan, just as they must budget. The general manager of the business unit carries the ultimate responsibility for the strategic plan, as for all other plans and results, but all functional executives should be involved in the process. The developed strategic plan (a *final* strategic plan seems like a contradiction in terms) should be presented to and discussed with the board of directors, if the firm constitutes a single operation, or with group management, if a division of a multiindustry company.

The planning process is iterative and messy.

How?   Strategic plans are not hatched; they are developed over time. Discussion and interchange—even friction and argument—are constructive and they need to occur up, down, and across in the organization. The process is iterative and messy; positions are taken and subsequently modified.

Thus, the process of strategy formulation (and reformulation) is typically fragmented and evolutionary. It may also be somewhat more intuitive than analytical. James Brian Quinn refers to the process of formulating and implementing strategic change as "logical incrementalism." He says, "strategic decisions do not lend themselves to aggregation into a massive decision matrix where all factors can be treated relatively simultaneously in order to arrive at a holistic optimum." Not only is the human mind unable to perform such a grand analysis, but the process of strategy setting takes time "to create awareness, build comfort levels, develop concensus, select and train people, and so forth."[5]

Eventually, concensus emerges, sometimes with some necessary and directed persuasion on the part of the general manager. The developed strategic plan must then be explicitly articulated—in writing—in order to gain the maximum communication benefit. The written document should spell out the consequences of the strategic plan on each of the functional areas of the business.

When? Plan continuously. High-technology companies operate, by definition, in very volatile and fast-changing environments. To refuse to adapt the business strategy on an interim basis is to court disaster. At the same time, strategic planning is fundamentally different from operational planning, and well thought out strategy will have a fair amount of permanency to it. To alter strategy too frequently is to negate the benefits that arise from experience, corporate culture, and clear perceptions of the company on the part of customers, investors, and suppliers. But failure to adapt strategy to meet changing economic and competitive conditions is far more dangerous.

Strategic planning must be continuous.

Any strategic plan deserves a thorough and careful re-evaluation at least annually. Boards of directors should insist on it. This annual review should not coincide with the annual budgeting process. When strategic planning and budgeting are attempted simultaneously, the strategic plan suffers neglect because operating managers become preoccupied with short-term planning that has immediate budgetary implications. Ideally, the strategic planning effort should occur about midway in the company's fiscal year: early enough to avoid confusion with the budgeting process, and yet late enough so that the need (or temptation) to redefine strategy will not occur as next year's profit plan is put together. Most companies develop a five-year (or longer) financial plan as a part of the strategic plan.

Facilitate discussion with an off-site planning meeting.

Where? Some of the preliminary skirmishing in connection with the annual strategic review needs to occur in short meetings around the office, across the lunch table, on airplanes, and over a friendly beer. These early encounters should stimulate the creative juices, solicit new ideas, and question conventional wisdoms. They are necessary precursors to a larger and longer meeting, preferably held off-site, in which the entire senior management group presents and discusses the preliminary strategic plan. The number involved is a matter of management style, but the group should probably include at least the level of management reporting to each functional executive. It is vital that the plan be considered preliminary at this point. If not, the off-site meeting becomes a defense, rather than a discussion, of the plan. The final written articulation of the plan should be forthcoming soon after the conclusion of the annual off-site planning conference.

Why Planning Fails

Without doubt, some business units are better strategic planners than others. What are some of the causes of poor strategic planning? There are scores of reasons, but some of the key shortcomings seem to fall into the following categories:

1. Excessive short-term orientation. Managers must be encouraged to think well beyond the current budget horizon. Focusing on plans two, three, and five years out may help overcome this short-term bias.

Look long term and fight against myopia.

2. Domination of a single functional viewpoint. Many companies exhibit a functional myopia and become known as "marketing-dominated" companies or "finance-dominated" companies. This organizational myopia often reflects that of the general manager; in turn, his or her myopia may be particularly strong if the general manager's previous experience was entirely within a single function. Job rota-

tion across functional lines for those managers destined for senior positions can mitigate this problem.

3. Unwillingness to grapple with the realities of *both* the market and the technology. This pitfall is obviously related to the previous one. When top managers are uncomfortable with the technologies that do or might shape the business—or when they are unwilling to devote time and attention to discovering customers' needs—an incomplete or suboptimal strategic plan is likely to result.

4. Failure to recognize multiple sources of strategic advantage. This pitfall is another type of myopia. The marketplace and the development laboratory are too frequently seen as the only sources of strategic advantage. Chapters 3 through 6 suggest ways in which manufacturing and finance can also contribute strategic leverage to the firm.

5. Restrained thinking. More generally, a management team may have difficulty seeing beyond the traditional boundaries of its own business. Many consulting firms have made a handsome business of assisting such firms. Individual facilitators or outside guests at strategic planning meetings may help dismantle some of these mental blocks. A well-informed and broadly experienced board of directors can offer much the same benefit to the company, if the management is willing to engage the board in the planning process.

6. Top management impatience that manifests itself in strategic plans that are not thoroughly articulated, incomplete communication of the plans, or lack of thorough understanding on the part of middle managers of the implications of the plans on their functions. Top managers must be communicators, teachers, and coaches. A business strategy that middle managers do not thoroughly understand and accept will be an incompletely implemented strategy.

Restrained thinking, impatience, and risk aversion are other common failings.

7. Excessive risk aversion. This problem is more prevalent in large than small companies. Managers of high-technology companies must see opportunities, as well as risks, in the rapidly changing environment and competitive conditions. The objective must not be to avoid the risks—for they are unavoidable—but rather to seize the opportunities. Chapter 8 suggests some steps that top management can take to guard against excessive risk aversion and to maintain creativity in the organization.

Risk and Opportunity

Remember that all risks present opportunities.

Throughout this book, I have emphasized the dynamic nature of high-technology industries, a dynamism that creates many opportunities while generating risks. Emergence of new market opportunities, evolving technology, and changing competitive structures are the constants with which all who are involved in strategy formulation must work. Change and evolution are the expected.

Table 9-1 presents a checklist of sources of dynamism within technology-based businesses, a brief explanation, and some examples of products or industries that have been particularly affected by the condition in recent years. Those engaged in strategic planning are well advised to spend some quiet time

Table 9-1. Sources of dynamism in high-technology industries

Source	Explanation	Examples
Technological uncertainty	Not only is new technology continually emerging, but an ongoing struggle occurs among available technologies for commercial dominance.	Semiconductor processing Video recording
Sales strategy uncertainty	The efficacy of alternative selling strategies is unclear when an industry is immature.	Personal computers Packaged software for business applications
High initial costs but steep experience curves	Scale economies, process improvements, shared experience, and vertical integration can rapidly alter industry economics.	High-speed computer printers Consumer electronic devices
Rapid changes in price-performance ratios	In essence, the combined effect of the first and third sources creates new buyer segments.	Telecommunication systems Home security systems
Emergence of entrepreneurial ventures	New niche market segments may attract fleet-footed and single-minded new companies before the segments are recognized by present suppliers.	Special-purpose computer systems Genetic engineering research

Table 9-1. (continued)

Source	Explanation	Examples
High costs of capital[a]	The competitive advantage will switch from small to large companies as an industry segment becomes capital intensive.[b]	"Catalog" semiconductors and integrated circuits High-volume computer peripherals
Market segment restructuring due to demographic changes, price & availability of substitute & complementary products, or saturation	The effect on high-technology companies may be second order: The primary impact is often on OEM customers.	Amateur photographic devices and supplies Color television
Increased buyer knowledge	As buyers become more discriminating, product standardization and heightened price competition result.	Personal computers 35mm cameras
Diffusion of proprietary knowledge	As knowledge spreads, competitive barriers built on proprietary technology must be rebuilt.[c]	Virtually all non-patented process technologies and new products
Changes in input cost, e.g., wages, material, exchange rates, capital, transportation, energy, communication	Factor costs in high-technology industries have experienced sharply different rates of gain and decline in recent years.	Petrochemical products Miniaturized electronic devices
Structural changes in adjacent industries	Related industries create both threats and opportunities.	Chemical & genetic engineering industries Solar & petroleum industries
Government policy changes (incentives & subsidies, controls and taxes, foreign trade restrictions, or defense policies)	Both home and foreign government actions can rapidly change an industry's "ground rules."	VLSI circuits (Japan & France) Solar energy Defense-sensitive equipment

[a]Note the advantage possessed by Japanese high-technology companies enjoying substantially lower costs of capital than their U.S. competitors.

[b]Consolidation through acquisition is likely to occur as an industry becomes more capital intensive.

[c]Knowledge spreads by means of technical journals and meetings, product specification sheets, employee mobility, and sometimes by illegal or unethical practices. This diffusion also contributes to the steep experience curve. (See the third source in this table.)

Source: Adapted from Michael Porter, *Competitive Strategy* (New York: Free Press, 1980), chap. 10 and pp. 163–184.

in carefully reviewing this checklist in an effort to ferret out those forces of change that are likely to demand some reformulation of the business's strategies within the next several years.

∴ Avoid the common pitfalls of poor strategic planning, particularly individual or organizational myopia. Use the checklist I present to sort through the risks and opportunities that arise from the rapid changes in technology, market needs, industry structures, and government policies that characterize high-technology industries. These changes inevitably force you to adapt or reform your business strategy from time to time.

Highlights

- Successful strategic planning in high-technology companies must involve all key functional managers, must assess both technological impacts and market needs, and requires seizing appropriate but risky opportunities.

- Your strategic plan should assure that your company delivers to customers an optimum mix of performance, packaging, support, quality/reliability, and price and earns above-average returns doing so.

- A dynamic process, strategic planning both drives and is driven by functional strategies and requires that market segment, sales, product, process, financial and human resources substrategies be well integrated.

- The strategy diagram assists in analyzing the implications and requirements of any business strategy, including the three generic strategies (low-cost producer, niche, and differentiation).

- Technological capabilities, technical leverage, market share, and market growth are the terms in which high-technology firms should assess their competitive positions.

- Changes in technology, market needs, industry structures, and government policies are the sources of risks and opportunities to which your evolving strategic plan must respond.

Notes

1. Peter F. Drucker, *Management: Tasks, Responsibilities, Practices* (New York: Harper & Row, 1973), pp. 123–125.

2. Ibid., p. 791.

3. In *In Search of Excellence* (New York: Harper & Row, 1982), Thomas J. Peters and Robert H. Waterman identify eight attributes of excellently managed companies. Most of these turn on people and organization strategy, including particularly bias for action, autonomy and entrepreneurship, productivity through people, and simple form, lean staff.

4. Michael Porter, *Competitive Strategy* (New York: Free Press, 1980), p. 39.

5. James Brian Quinn, "Strategic Change: 'Logical Incrementalism,'" *Sloan Management Review* (Fall 1978), pp. 7–17.

Index